Study Guide for Moore's

Basic Practice of Statistics
Second Edition

William I. Notz
Michael A. Fligner
Ohio State University

Rebecca Sorice

W. H. Freeman and Company
New York

Copyright © 2000 by W. H. Freeman and Company

Printed in the United States of America

ISBN: 0-7167-3617-9

First printing 1999

CONTENTS

PREFACE

We have written this Study Guide to help you learn and review the material in the text. The structure of this study guide is as follows. We provide an *overview* of each section that reviews the key concepts of the section. After the overview, there are *guided solutions to selected problems* in that section, along with the *key concepts* from the section required for solving each problem. The guided solutions provide hints for setting up and thinking about the exercise, which should help improve your problem-solving skills. Once you have worked through the guided solution, you can look up the *complete solution* provided later in the study guide to check your work.

What is the best way to use this study guide? Part of learning involves doing homework problems. In doing a problem, it is best for you to first try to solve the problem on your own. If you are having difficulties, then try to solve the problem using the hints and ideas in the guided solution. If you are still having trouble, read through the complete solution. Generally, try to use the complete solution as a way to check your work. Be careful not to confuse reading the complete solutions with doing the problems themselves. This is the same mistake as reading a book about swimming and then believing you are prepared to jump into the deepest end of the pool. If you simply read our complete solutions and convince yourself that you could have worked the problem on your own, you may be misleading yourself and could have trouble on exams. When you are having difficulty with a particular type of problem, find similar problems to work in the exercises in the text. Often problems adjacent to each other in the text use the same ideas.

In the overviews we try to summarize the material that is most important, which should help you review material when you are preparing for a test. If any of the terms in the overview are unfamiliar, you probably need to go back to the text and reread the appropriate section. If you are having difficulties with the material, don't neglect to see your instructor for help. Face to face communication with your instructor is the best way to clear up difficulties.

CHAPTER 1

EXAMINING DISTRIBUTIONS

SECTION 1.1

OVERVIEW

Understanding data is one of the basic goals in statistics. To begin, identify the **individuals** or objects described, then the **variables** or characteristics being measured. Once the variables are identified, you need to determine whether they are **categorical** (the variable puts individuals into one of several groups) or **quantitative** (the variable takes meaningful numerical values for which arithmetic operations make sense). The Guided Solution for Exercise 1.11 gives more detail on deciding whether a variable is categorical or quantitative.

After looking over the data and digesting the story behind it, the next step is to describe the data with graphs. The first methods are simple graphs that give an overall sense of the pattern of the data. Which graphs are appropriate depends on whether the data are numerical or not. Categorical data (nonnumerical data) use **bar charts** or **pie charts**. Quantitative data (numerical data) use **histograms** or **stemplots**. Quantitative data collected over time use a **timeplot** in addition to a histogram or stemplot.

When examining graphs be on the alert for

• **outliers** (unusual values) that do not follow the pattern of the rest of the data

- some sense of a **center** or typical value of the data

- some sense of how **spread** out or variable the data are

- some sense of the **shape** of the **overall pattern**

In timeplots, be on the lookout for **trends** over time. These features are important whether we draw the graphs ourselves or depend on a computer to draw them for us.

GUIDED SOLUTIONS

Exercise 1.11

KEY CONCEPTS - individuals and type of variables

a) When identifying the individuals or objects described, you need to include sufficient detail so that it is clear which individuals are contained in the data set.

b) Recall that the variables are the characteristics of the individuals. Once the variables are identified, you need to determine if they are categorical (the variable just puts individuals into one of several groups) or quantitative (the variable takes meaningful numerical values for which arithmetic operations make sense). Note that if we had another variable, "League," which was scored as a 1 if the player was in the American League and a 0 if the player was in the National League, this would still be a categorical variable even though we used numbers to represent the two categories. Now list the variables recorded and classify each as categorical or quantitative.

<u>Name of variable</u>   <u>Type of variable</u>

c) Be careful with the units for salary.

Exercise 1.15

KEY CONCEPTS - describing histograms: shape and center

a) Look at the overall pattern without focusing on minor irregularities (small ups and downs in the heights of the bars). Find the approximate center and see if the right and left sides seem similar (symmetric) or if one side tends to extend farther out than the other (skewed to the right or left).

b) The batting average of a typical player is described by the center of the histogram. This is the value where about half the observations have smaller values and the other half have larger values. You can form a pretty accurate guess from visual inspection of the histogram or you can use the following more formal approach. There are 167 players in the dataset, so you need to find a number with about 167/2 or approximately 83 observations below it. This can be done by adding up the heights of the histogram bars starting at the left. It will not be possible to get a sum exactly equal to 83, but this method will allow you to determine which interval contains the center. This is the best we can do with the information provided in a histogram. (Note that in a stemplot the values of all the observations are given so we could find the central observation exactly.)

    Interval containining the center:
    Approximate center:

Unlike a stemplot, we cannot determine the minimum and maximum exactly from a histogram. However, we can approximate their values. To approximate the highest and lowest batting averages, first find the bar containing the highest and lowest batting averages (excluding George Brett). We know that the lowest batting average is in the interval .185 to .195 and that there is only one batting average in this interval. With no further information, we could use the midpoint of this interval as an approximate value for the lowest batting average or simply indicate that the lowest batting average is between .185 and .195.

Exercise 1.17

KEY CONCEPTS - drawing a histogram, comparing two histograms

Hints to remember in drawing a histogram:
1. Divide the range of values of the data into classes or intervals of equal length. (Notice that the data in Table 1.3 are already this way.)

2. Count the number of data values that fall into each interval. (Again, this step is done in Table 1.3.)

3. Draw the histogram.

 a. Mark the intervals on the horizontal axis and label the axis. Include the units.

 b. Mark the scale for the counts or percents on the vertical axis. Label the axis.

 c. Draw bars, centered over each interval, up to the height equal to the count or percent. There should be no space between the bars (unless the count for a class is zero, which creates a space between bars).

For the exercise:
a) The vertical axis in a histogram can be either the count or percent of the data in each interval. Changing the units from counts to percents will not affect the shape of the histogram. However, using percents makes it simpler to compare two histograms based on different numbers of observations, by making the scales comparable. Complete the table of the percent of the total population in each age group. The entry 19.4 given in the table was obtained by dividing 29.3 million, the number under 10 years in 1950, by 151.1 million, the total population in 1950.

Age Distribution (Percentages), 1950 and 2075

Age group	1950	2075
Under 10 years	19.4	
10 to 19 years		
20 to 29 years		
30 to 39 years		
40 to 49 years		
50 to 59 years		
60 to 69 years		
70 to 79 years		
80 to 89 years		
90 to 99 years		
100 to 109 years		
Total		

b, c) Complete the two histograms given. The values of the variable year have gaps. To avoid the gaps in the histograms, the bases of the bars need to be extended to meet halfway between two adjacent values. For example, the bar representing 30 - 39 years would need to meet the bar from 20 - 29 years at 29.5 Similarly, the bar representing 30 - 39 years would need to meet the bar from 40 - 49 years at 39.5. Thus the bar representing 30 - 39 years must go from 29.5 to 39.5, with a center at 34.5. In the histograms given, the *centers* of the bars are provided on the horizontal axis, and the first bar for the 1950 age distribution has been completed for you.

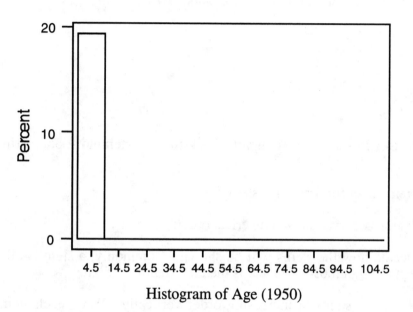

Histogram of Age (1950)

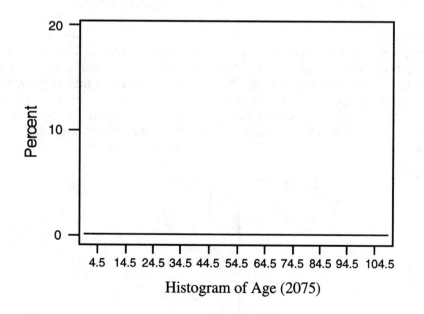

Histogram of Age (2075)

Describe the shapes of the two histograms. Are they approximately symmetric or are they skewed? What are their centers, and what can you say about the spread of each? By comparing the two histograms, what are the most important changes in age distribution between 1950 and 2075?

Exercise 1.19

KEY CONCEPTS - drawing stemplots, back-to-back stemplots, comparing two stemplots

Hints to remember for drawing a stemplot:

1. Put the observations in numerical order.

2. Decide how the stems will be shown. Commonly, a stem is all digits except the rightmost. The leaf is then the rightmost digit.

3. Write the stems in increasing order vertically. Write each stem only once. Draw a vertical line next to the stems.

4. Write each leaf next to its stem.

5. Rewrite the stems and put the leaves in increasing order.

For the exercise:
We have given the stems below, enclosed in vertical lines. The stems are in units of 10 home runs. Make the stemplot for Ruth in the usual way to the right of the stems. The stemplot for McGwire is made in the same way, with the leaves going off to the left instead of the right. This type of plot is excellent for comparing two small data sets. Complete the back-to-back stemplot below. Two values for McGwire and one value for Ruth have been included to help you get started.

```
McGwire        Ruth
         99 | 0 |
            | 1 |
            | 2 | 2
            | 3 |
            | 4 |
            | 5 |
            | 6 |
            | 7 |
```

How do the two players compare? Concentrate on center, spread, outliers and overall shape of the two stemplots.

Exercise 1.21

KEY CONCEPTS - drawing and interpreting a timeplot

a) Complete the timeplot on the graph. The winning times for 1972 and 1973 are included in the plot to get you started.

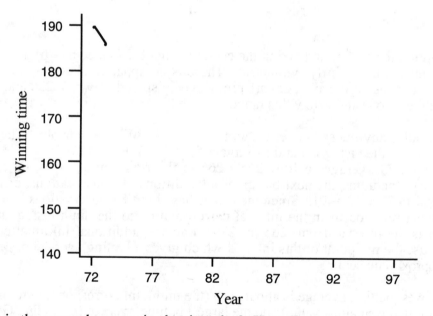

What is the general pattern in the timeplot? Have times stopped improving in recent years?

COMPLETE SOLUTIONS

Exercise 1.11

a) The individuals are all major league baseball players as of opening day of the 1998 season.

b) There are four variables in addition to the player's name. They are listed below.

Name of variable	Type of variable
Team	Categorical
Position	Categorical
Age	Quantitative
Salary	Quantitative

c) Age is in years, and salary is in thousands of dollars. If you gave the units of salary as dollars, it would imply that the salary for Eduardo Perez was $300 per year.

Exercise 1.15

a) Ignoring the outlier and taking the center at about .270 (see part b), the shape of the histogram is fairly symmetric. There is no apparent skewness in either direction. Most of the averages are pretty evenly spread between .230 and .300, with a few above and below this range.

b) Visually, any guess between about .260 and .280 is reasonable. The first eight bars of the histogram add to a total of $1 + 3 + 1 + 4 + 13 + 20 + 15 + 16 = 73$, giving 73 averages below .265 (.265 is the right endpoint of the eighth interval). Including the next bar gives an additional 18 observations, bringing the total to $73 + 18 = 91$. Since the center has about 83 observations below it, the center must occur in the interval corresponding to the ninth histogram bar which is the interval from .265 to .275. Lacking additional information, we could use the midpoint of this interval which gives a batting average of .270 as the approximate center.

The lowest batting average is about .190 (the midpoint corresponding to the bar containing the smallest value) and the largest batting average (excluding George Brett) is about .350.

Exercise 1.17

Age group	1950	2075
Under 10 years	19.4	11.2
10 to 19 years	14.4	11.5
20 to 29 years	15.9	11.8
30 to 39 years	15.1	12.3
40 to 49 years	12.8	12.2
50 to 59 years	10.3	12.1
60 to 69 years	7.3	11.1
70 to 79 years	3.6	8.8
80 to 89 years	1.1	6.1
90 to 99 years	0.1	2.5
100 to 109 years		0.5
Total	100.0	100.1

b, c) The histograms are reproduced below. They use the same numerical scale as in the guided solutions, but the numbers have been suppressed.

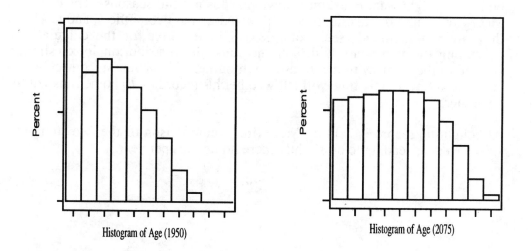

Histogram of Age (1950)   Histogram of Age (2075)

The center of the 1950 age distribution is pretty close to 30 years - there are 49.7% of the people with ages up to 29 (the first three age intervals) with the remaining 50.3% above. In this case, the calculation is most easily done by adding the percentages given in part a rather than trying to read them off the histogram. For the projected 2075 age distribution, the center is in the 40 - 49 year age interval, probably close to the midpoint, or about 44 years. We can't specify it more exactly from the information given. The 1950 age distribution is skewed to the right and has much less spread than the projected 2075 age distribution. The shape for the projected 2075 age distribution has the ages pretty evenly spread up to 70 or 80 years (the histogram is pretty flat until this point because the percents are similar for these intervals), with a small percentage (just under 10%) exceeding 80 years.

Exercise 1.19

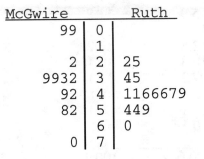

```
 McGwire          Ruth
      99 | 0 |
         | 1 |
       2 | 2 | 25
    9932 | 3 | 45
      92 | 4 | 1166679
      82 | 5 | 449
         | 6 | 0
       0 | 7 |
```

Ruth's distribution is centered at 46 home runs per year and looks fairly symmetric. His famous record of 60 home runs in 1927 does not appear to be an outlier, given his record.

The years in which McGwire was injured and the baseball strike each have 9 home runs. While they show up as outliers, there are good explanations for both. We might want to exclude these years as not full seasons. The center of McGwire's home run distribution is 39 including the two outliers, and is 41.5 if they are not included (see Example 1.8 in the text for the calculation). Excluding the two years with 9 home runs, the distribution looks slightly skewed to the right, with the 70 as a high outlier. However, since McGwire's career is not over, only time will tell whether his record of 70 home runs will be an outlier.

Although the record is held by McGwire, a comparison of the two stemplots shows Ruth as tending to hit slightly more home runs per year.

Exercise 1.21

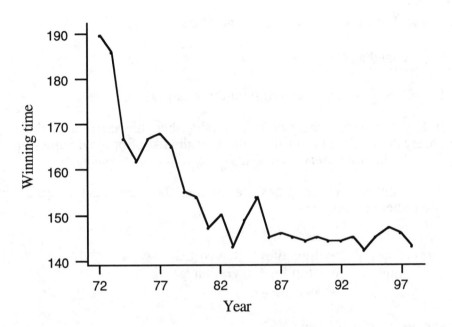

b) There is a fairly steady downward trend in the winning times until around 1980. After this point the improvement seems to have stopped with the winning times varying around 145 minutes. The variability in the winning times from year to year has also gone down.

SECTION 1.2

OVERVIEW

Once you examine graphs to get an overall sense of the data, it is helpful to look at numerical summaries of features of the data that clarify the notions of center and spread.

Measures of center: • **mean** (often written as $\bar{x}$)
                     • **median** (often written as M)

Finding the mean $\bar{x}$

The mean is the common arithmetic average.

If there are n observations, $x_1, x_2, \ldots, x_n$, then the mean is

$$\bar{x} = \frac{x_1 + x_2 + \ldots + x_n}{n} = \frac{1}{n}\sum x_i$$

Recall that $\sum$ means "add up all these numbers."

<u>Finding the median M</u>

1. List all the observations from smallest to largest.

2. If the number of observations is odd, then the median is the middle observation. Count from the bottom of the list of ordered values up to the $(n + 1)/2$ largest observation. This observation is the median.

3. If the number of observations is even, then the median is the average of the two center observations.

Measures of spread: • **quartiles** (often written as Q_1 and Q_3)
  • **standard deviation** (s)
  • **variance** (s^2)

<u>Finding the quartiles Q_1 and Q_3</u>

1. Locate the median.

2. The first quartile, Q_1, is the median of the lower half of the list of ordered observations.

3. The third quartile, Q_3, is the median of the upper half of the list of ordered values.

<u>Finding the variance s^2 and the standard deviation s</u>

1. Take the average of the squared deviations of each observation from the mean. In symbols, if we have n observations, $x_1, x_2, \ldots, x_n$ with mean $\bar{x}$

$$s^2 = \frac{(x_1 - \bar{x})^2 + (x_2 - \bar{x})^2 + \ldots + (x_n - \bar{x})^2}{n - 1} = \frac{1}{n-1}\sum(x_i - \bar{x})^2$$

(Remember, $\sum$ means "add up.")

If you are doing this calculation by hand, it is best to take it one step at a time. First calculate the deviations, then square them, next sum them up, and finally divide the result by $n - 1$.

2.   The standard deviation is the square root of the variance, i.e., $s = \sqrt{s^2}$. Some things to remember about the standard deviation:

a. s measures the spread around the mean.

b. s should be used only with the mean, not with the median.

c. If $s = 0$, then all the observations must be equal.

d. The larger s is, the more spread out the data are.

e. s can be strongly influenced by outliers. It is best to use s and the mean only if the distribution is symmetric or nearly symmetric.

For measures of spread, the quartiles are appropriate when the median is used as a measure of center. In fact, the **five-number summary**, reporting the largest and smallest values of the data, the quartiles, and the median, provides a compact description of the data. The five-number summary can be represented graphically by a **boxplot**. If you use the mean as a measure of center, then the standard deviation and variance are the appropriate measures of spread. Watch out, because means and variances can be strongly affected by outliers and are harder to interpret for skewed data. The mean and standard deviation are not resistant measures. The median and quartiles are more appropriate when outliers are present or when the data are skewed. The median and quartiles are **resistant measures**.

GUIDED SOLUTIONS

Exercise 1.37

KEY CONCEPTS - measures of center: mean and median

The rule for finding the mean is to add the values of all the observations and then divide by the number of observations. How many observations (employees) are there? What is the total of their salaries?

    Number of observations   =
    Total of the observations   =
    Mean of the observations   =

How many salaries are less than the mean? Why do so many of the employees make less than the mean?

Remember, to find the median, first order the observations, making sure to include a salary as many times as it occurs in the data set. If the number of observations is odd, the median is the observation in the center. If the number of observations is even, the median is the average of the two center observations.

Exercise 1.39

KEY CONCEPTS - boxplots, side-by-side boxplots

Hints for drawing a boxplot:

1. The center box starts at Q_1 and the ends at Q_3.

2. The median is marked by a line in the center of the box.

3. Lines extend out from the box to the smallest and largest observations.

The boxplot for the beef hot dogs follows. Complete the side-by-side boxplots by drawing the boxplots for meat and poultry. All you need to draw a boxplot is the five-number summary, and this information is contained in the computer output provided.

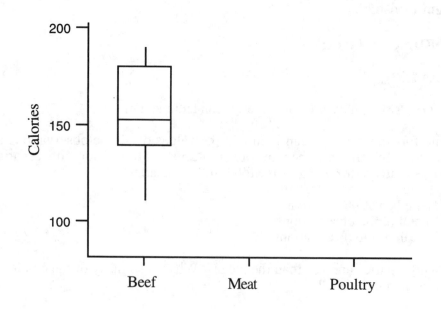

How do the three distributions compare? Will eating poultry hot dogs usually lower your calorie consumption compared to beef or meat?

Exercise 1.41

KEY CONCEPTS - histograms, stemplots and boxplots, mean and standard deviation

The simplest graph to start with when you have a small data set is a stemplot. The first two stems are given below. Fill in the rest of the stems and complete the stemplot. See Exercise 1.19 for hints on how to make a stemplot.

```
4.8 | 8
4.9 |
    |
    |
    |
    |
    |
    |
    |
    |
```

Given the distribution of measurements, are $\bar{x}$ and s good measures to describe this distribution? Find their values and give an estimate of the density of the earth based on these measures. Remember, if you are doing the calculation by hand, first find $\bar{x}$ and then calculate s in steps.

COMPLETE SOLUTIONS

Exercise 1.37

There are five clerks, two junior accountants, and the owner for a total of eight employees. The total of the salaries is

Total salary of clerks   5 x $22,000 = $110,000
Total salary of accountants   2 x $50,000 = $100,000
Salary of owner   = $270,000
Total of all salaries   = $480,000

Number of salaries   8
Mean of the observations $480,000/8 = $60,000

Everyone but the owner makes less than the mean. This is because the mean is not a resistant measure of center and its value is pulled up by the high "outlier" (very large salary of the owner).

The ordered observations are

$22,000 $22,000 $22,000 $22,000 $22,000 $50,000 $50,000 $270,000

The median is the average of the fourth and fifth observations which is $22,000.

Exercise 1.39

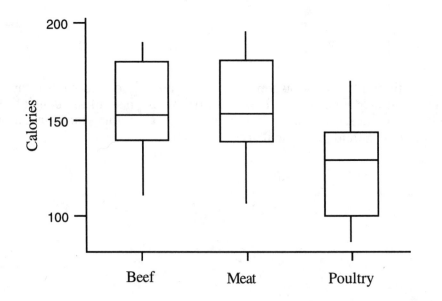

A comparison of the boxplots for beef and meat shows little difference in these two distributions. However, poultry hot dogs are generally lower in calories than either beef or meat. The lower quartile of the poultry distribution is below the minimum number of calories for beef or meat, indicating that about one fourth of the poultry brands included in the study had fewer calories than any of the beef or meat brands included in the study.

Exercise 1.41

The stemplot gives a distribution that appears fairly symmetric with one outlier of 4.88, but this is not that far from the bulk of the data. In this case, $\bar{x}$ and s should provide reasonable measures of center and spread.

```
4.8 | 8
4.9 |
5.0 | 7
5.1 | 0
5.2 | 6799
5.3 | 04469
5.4 | 2467
5.5 | 03578
5.6 | 12358
5.7 | 59
5.8 | 5
```

In practice, you will be using software or your calculator to obtain the mean and standard deviation from keyed-in data. However, we illustrate the step-by-step calculations, to help you understand how the standard deviation works. Be careful not to round off the numbers until the last step, as this can sometimes introduce fairly large errors when computing s.

Observation	Difference	Difference squared
x_i	$x_i - \bar{x}$	$(x_i - \bar{x})^2$
5.50	0.052100	0.002714
5.61	0.162100	0.026277
4.88	-0.567900	0.322510
5.07	-0.377900	0.142808
5.26	-0.187900	0.035306
5.55	0.102100	0.010424
5.36	-0.087900	0.007726
5.29	-0.157900	0.024932
5.58	0.132100	0.017450
5.65	0.202100	0.040845
5.57	0.122100	0.014908
5.53	0.082100	0.006740
5.62	0.172100	0.029618
5.29	-0.157900	0.024932

Observation	Difference	Difference squared (cont.)
x_i	$x_i - \bar{x}$	$(x_i - \bar{x})^2$
5.44	-0.007900	0.000062
5.34	-0.107900	0.011642
5.79	0.342100	0.117033
5.10	-0.347900	0.121034
5.27	-0.177900	0.031648
5.39	-0.057900	0.003352
5.42	-0.027900	0.000778
5.47	0.022100	0.000488
5.63	0.182100	0.033161
5.34	-0.107900	0.011642
5.46	0.012100	0.000146
5.30	-0.147900	0.021874
5.75	0.302100	0.091265
5.68	0.232100	0.053870
5.85	0.402100	0.161684
Column sums = 157.99		1.366869

The mean is 157.99/29 = 5.4479. This has been subtracted from each observation to give the second column of deviations or differences $x_i - \bar{x}$. The second column is squared to give the squared deviations or $(x_i - \bar{x})^2$ in the third column. The variance is the sum of these squared deviations divided by one less than the number of observations:

$$s^2 = \frac{1.366869}{29-1} = .0488168 \text{ and } s = \sqrt{.0488168} = .22095$$

Use the mean of 5.4479 as the estimate of the density of the earth based on these measurements, since the distribution is approximately symmetric, and without outliers.

SECTION 1.3

OVERVIEW

Section 1.3 considers the use of **mathematical models** (mathematical formulas) to describe the overall pattern of a distribution. The name given to a mathematical model that summarizes the shape of a histogram is a **density curve**. The density curve is an idealized histogram. The area under a density curve between two numbers represents the proportion of the data that lie between these two numbers. Like a histogram, it can be described by measures of center such as the **median** (a point such that half the area under the density curve is to the left of the point) and the **mean** (the center of gravity or balance point of the density curve). In Section 1.2 we called the mean $\bar{x}$. This is how to refer to the mean of actual observations. The mean of a density curve is referred to as μ. Likewise, the standard deviation of a density curve also has a new notation. It is referred to as σ.

One of the most commonly used density curves in statistics is the **normal curve**. Normal curves are symmetric and bell-shaped. The peak of the curve is located above the mean and median, which are equal since the density curve is symmetric. The standard deviation measures how concentrated the area is around this peak. Normal curves follow the 68 - 95 - 99.7 rule, i.e., 68% of the area under a normal curve lies within one standard deviation of the mean (illustrated in the figure), 95% within two standard deviations of the mean, and 99.7% within three standard deviations of the mean.

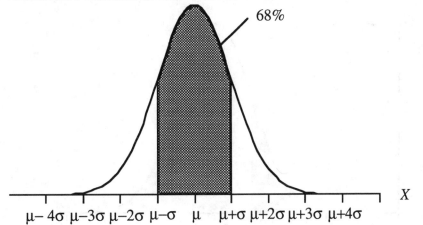

Areas under any normal curve can be found easily if quantities (x) are first standardized by subtracting the mean (μ) from each value and dividing the result by the standard deviation (σ). This **standardized value** is called the *z-score*.

$$z = \frac{x - \mu}{\sigma}$$

If data whose distribution can be described by a normal curve are standardized (all values replaced by their z-scores), the distribution of these standardized values is described by the **standard normal curve**. Areas under standard normal curves are easily computed by using a **standard normal table** such as Table A on the inside front cover of the text.

The standard normal curve is very useful for finding the proportion of observations in an interval when dealing with any normal distribution. Here are some hints about solving these problems:

1. State the problem.

2. Draw a picture of the problem. It will help you know which area you are looking for.

3. Standardize the observations.

4. Using Table A from the text, find the area you need.

Hint: The normal curve has a total area of 1. The normal curve is also symmetric so areas (proportions) such as those shown in the next two figures are equal.

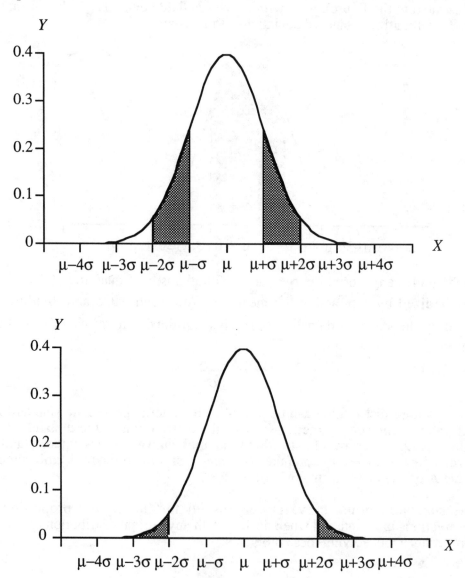

GUIDED SOLUTIONS

Exercise 1.63

KEY CONCEPTS - the 68-95-99.7 rule for normal density curves

a) Recall that the 68-95-99.7 rule states

68% of the data will be between $\mu - \sigma$ and $\mu + \sigma$.
95% of the data will be between $\mu - 2\sigma$ and $\mu + 2\sigma$.
99.7% of the data will be between $\mu - 3\sigma$ and $\mu + 3\sigma$.

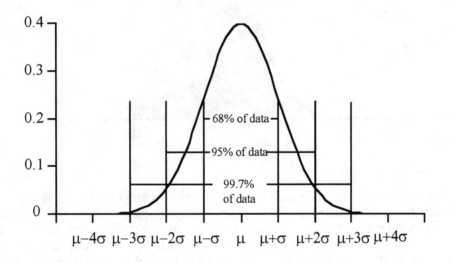

What are μ and σ in this problem? Now use this to find the values between which 95% of all pregnancies fall.

b) Refer to the figure in part a. Below what value in the figure does 2.5% of the data fall? It is one of the values $\mu - 3\sigma$, $\mu - 2\sigma$, or $\mu - \sigma$. Remember the normal curve is symmetric about μ.

Now convert this to a value in days.

Exercise 1.65

KEY CONCEPTS - computing areas under a standard normal density curve

Recall that the proportion of observations from a standard normal distribution that are less than a given value z is equal to the area under the standard normal curve to the left of z. Table A gives these areas. This is illustrated in the following figure.

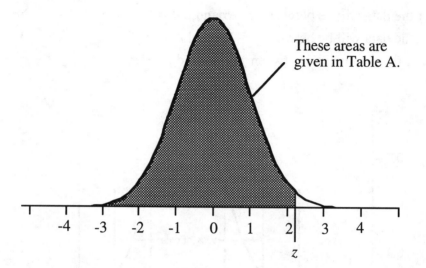

These areas are given in Table A.

In answering questions concerning the proportion of observations from a standard normal distribution that satisfy some relation, we find it helpful to first draw a diagram of the area under a normal curve corresponding to the relation. We then try to visualize this area as a combination of areas of the form in the previous figure, since such areas can be found in Table A. The entries in Table A are then combined to give the area corresponding to the relation of interest.

This approach is illustrated in the solutions that follow.

a) To get you started, we will work through a complete solution. A diagram of the desired area follows.

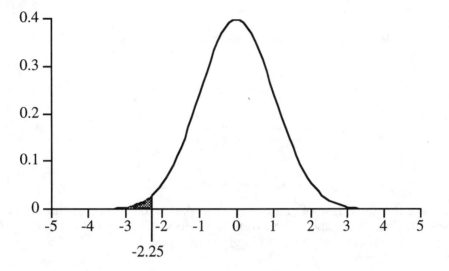

This is exactly the type of area that is given in Table A. We simply find the row labeled -2.2 along the left margin of the table, locate the column labeled .05 across the top of the table, and read the entry in the intersection of this row and column. We find this entry is 0.0122. This is the proportion of observations from a standard normal distribution that satisfies $z < -2.25$

b) Shade the desired area in the following figure.

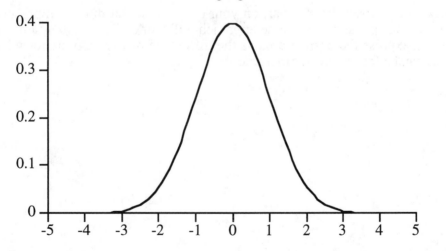

Remembering that the area under the whole curve is 1, how would you modify your answer from part a?

Area =

c) Try solving this part on your own. To begin, draw a diagram of a normal curve and shade the region.

Now use the same line of reasoning as in part b to determine the area of your shaded region. Remember, you want to try to visualize your shaded region as a combination of areas of the form in given in Table A.

d) To test yourself, try this part on your own. It is a bit more complicated than the previous parts, but the same approach will work. Draw a diagram and then try and express the desired area as the difference of two regions for which the areas can be found directly in Table A.

Exercise 1.67

KEY CONCEPTS - computing the area under an arbitrary normal density curve, finding the value x (the quantile) corresponding to a given area under an arbitrary normal density curve

a) What does the 68-95-99.7 rule tell us about the range in which the middle 95% of all yearly returns lie? You will need to identify μ and σ to apply the rule.

b) For this and the next part, we must first convert the question into one about a standard normal. This involves standardizing the numerical conditions by subtracting the mean and dividing the result by the standard deviation. We then draw a picture of the desired area corresponding to these standardized conditions and compute the area as we did for the standard normal, using Table A.

To carry this out, we must convert $x = 0\%$ to a z-score. Compute the z-score by completing the following. Recall from part a what μ and σ are.

$$z\text{-score} = \frac{x - \mu}{\sigma} =$$

The market is down if the return is less than 0. In terms of the standard normal, this is equivalent to determining the area under the standard normal curve that is below the z-score you computed above. Use Table A to compute this area. You may find it helpful to draw a picture of the desired area in the space provided. Refer to Exercise 1.65 if you need to review how to compute areas under the standard normal curve.

The area just found is the desired proportion.

c) Try this part on your own. Compute the appropriate z-score, determine what the question is asking in terms of an area under a standard normal curve, and compute this area.

Exercise 1.69

KEY CONCEPTS - computing the area under an arbitrary normal density curve

a) As in Exercise 1.67b and c, we must first convert the question into one about a standard normal. This involves standardizing the numerical conditions by subtracting the mean and dividing the result by the standard deviation. We then draw a diagram of the desired area corresponding to these standardized conditions and compute the area as we did for the standard normal, using Table A.

Try to carry this out on your own. Refer to exercise 1.67 if you need to refresh your memory concerning the details. Remember to express your answer as a percent.

b) This is similar to part a, but you must now use the mean for present-day children, not the mean for children in 1932.

COMPLETE SOLUTIONS

Exercise 1.63

a) In this problem the mean is $\mu = 266$ days and the standard deviation is $\sigma = 16$ days. From the 68-95-99.7 rule we know that the middle 95% of all pregnancies should fall between

$$\mu - 2\sigma = 266 - (2 \times 16) = 266 - 32 = 234 \text{ days}$$

and

$$\mu + 2\sigma = 266 + (2 \times 16) = 266 + 32 = 298 \text{ days}$$

b) Since 95% of all pregnancies fall between $\mu - 2\sigma = 234$ days and $\mu + 2\sigma = 298$ days, the remaining 5% of all pregnancies should take less than 234 days or more than 298 days. The symmetry of the normal curve about its mean implies that half of this 5% (in other words, 2.5%) will be below 234 days and the remaining 2.5% above 298 days. Thus the shortest 2.5% of all pregnancies are less than 234 days.

Exercise 1.65

a) A complete solution is provided in the guided solutions.

b) The desired area is indicated in the figure.

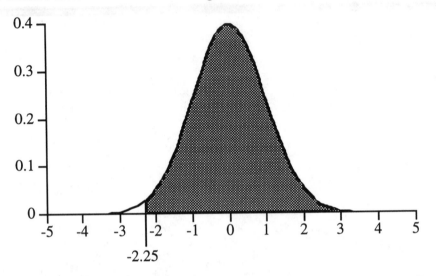

This is not of the form for which Table A can be used directly. However, the unshaded area to the left of -2.25 is of the form needed for Table A. In fact, we found the area of the unshaded portion in part a. We notice that the shaded area can be visualized as what is left after deleting the unshaded area from the total area under the normal curve.

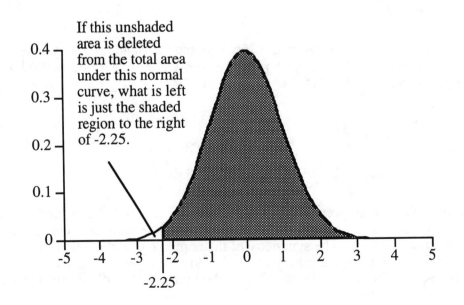

If this unshaded area is deleted from the total area under this normal curve, what is left is just the shaded region to the right of -2.25.

Since the total area under a normal curve is 1, we have

Shaded area = total area under normal curve - area of unshaded portion
= 1 - 0.0122 = 0.9878

Thus the desired proportion is 0.9878.

c) The desired area is indicated in the figure.

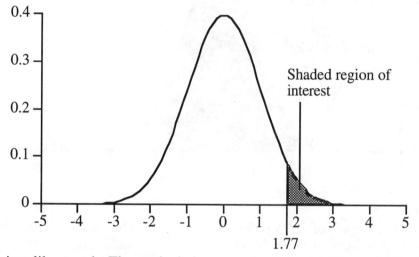

This is just like part b. The unshaded area to the left of 1.77 can be found in Table A and is 0.9616. Thus

Shaded area = total area under normal curve - area of unshaded portion
= 1 - 0.9616 = 0.0384.

This is the desired proportion.

d) We begin with a diagram of the desired area.

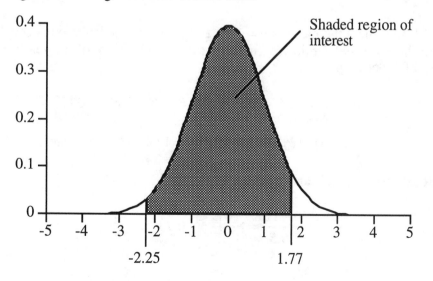

The shaded region is a bit more complicated than in the previous parts, but the same strategy still works. We note that the shaded region is obtained by removing the area to the left of -2.25 from all the area to the left of 1.77.

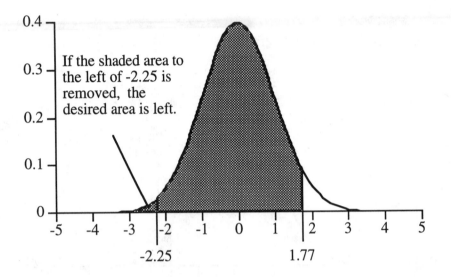

The area to the left of -2.25 is found in Table A to be 0.0122. The area to the left of 1.77 is found in Table A to be 0.9616. The shaded area is thus

$$\text{Shaded area} = \text{area to left of 1.77 - area to left of -2.25}$$
$$= 0.9616 - 0.0122$$
$$= 0.9494.$$

This is the desired proportion.

Exercise 1.67

a) In this problem the mean is $\mu = 12\%$ and the standard deviation is $\sigma = 16.5\%$. From the 68-95-99.7 rule we know that the middle 95% of all yearly returns should fall between

$$\mu - 2\sigma = 12\% - (2 \times 16.5\%) = 12\% - 33\% = -21\%$$

and

$$\mu + 2\sigma = 12\% + (2 \times 16.5\%) = 12\% + 33\% = 45\%$$

b) Recall that $\mu = 12\%$ and $\sigma = 16.5\%$, so

$$z\text{-score} = \frac{x - \mu}{\sigma} = \frac{0 - 12}{16.5} = \frac{-12}{16.5} = -0.73$$

The desired area is the area under the standard normal to the left of -0.73 as in the figure.

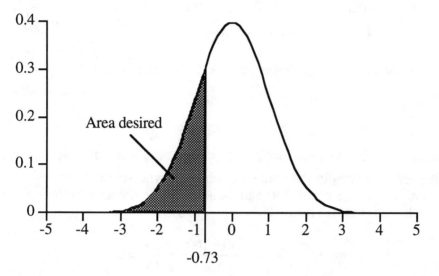

The area to the left of -0.73 can be read directly from Table A and is 0.2327. Thus the proportion of years the market is down is 0.2327.

c) Again, $\mu = 12\%$ and $\sigma = 16.5\%$ so

$$z\text{-score} = \frac{x - \mu}{\sigma} = \frac{25 - 12}{16.5} = \frac{13}{16.5} = 0.79$$

The desired area is the area under the standard normal to the right of 0.79, as in the figure.

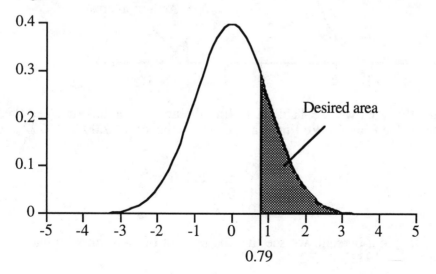

The area to the right of 0.79 is computed by first finding the area to the left of 0.79 in Table A and then subtracting this area from 1. We find the area to the left of 0.79 to be 0.7852. The area to the right of 0.79 is therefore

$$1 - 0.7852 = 0.2148$$

Thus the proportion of years the index gains 25% or more is 0.2148.

Exercise 1.69

a) For children in 1932 the mean score is $\mu = 100$ and the standard deviation is $\sigma = 15$. We are interested in the percent of children who score above 130. We must compute the z-score of $x = 130$ and then determine the area to the right of this value under the standard normal:

$$z\text{-score} = \frac{x - \mu}{\sigma} = \frac{130 - 100}{15} = \frac{30}{15} = 2$$

A diagram of the desired area under the standard normal curve follows.

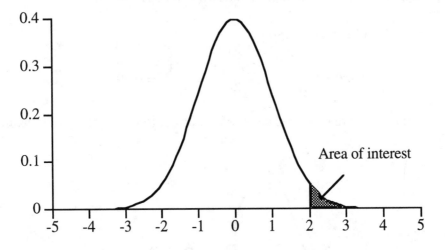

To compute this area we use Table A to find the area to the left of 2 and then subtract this value from 1. We find that the area to the left of 2.00 is 0.9772 and hence

$$\text{Desired area} = 1 - \text{area to left of 2}$$
$$= 1 - 0.9772$$
$$= 0.0228$$

Converting this to a percent, we see that the percent of children who had very superior scores in 1932 was 2.28%.

b) We proceed as in part a, but now we have $\mu = 120$ and $\sigma = 15$. We are again interested in the percent of children who score above 130. We must compute the z-score of $x = 130$ and then determine the area to the right of this value under the standard normal.

$$z\text{-score} = \frac{x - \mu}{\sigma} = \frac{130 - 120}{15} = \frac{10}{15} = 0.67$$

A picture of the desired area under the standard normal curve follows.

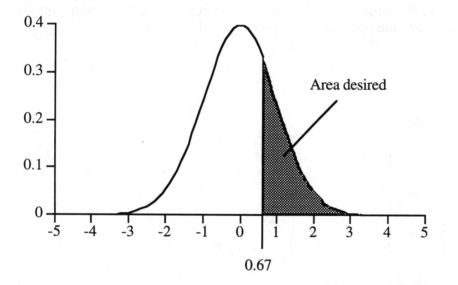

0.67

To compute the area desired we use Table A to find the area to the left of 0.67 and then subtract this value from 1. We find that the area to the left of 0.67 is 0.7486 and hence

$$\begin{aligned} \text{Desired area} &= 1 \text{ - area to left of } 0.67 \\ &= 1 - 0.7486 \\ &= 0.2514 \end{aligned}$$

Converting this to a percent, we see that the percent of present-day children who would have had very superior scores in 1932 is 25.14%.

SELECTED TEXT REVIEW EXERCISES

GUIDED SOLUTIONS

Exercise 1.73

KEY CONCEPTS - bar charts, pie charts

These data can be displayed in either a bar chart or a pie chart. Complete the following bar chart. Why do you need the "other methods" category? Make sure you can draw the pie chart as well. The complete solution contains both the bar chart and the pie chart corresponding to these data.

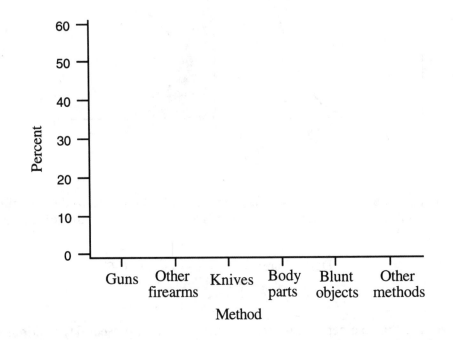

Exercise 1.79

KEY CONCEPTS - stemplots, boxplots, five-number summary, means and medians.

a) The simplest plot for a small data set such as this is a stemplot. If you used only two stems, one for the mileages in the 10s and one for those in the 20s, most of the features of the distribution would be obscured. In this case splitting the stems is helpful: split each of these stems into five stems. The first stem starting with 1 would have leaves 1 and 2, the second stem starting with 1 would have leaves 3 and 4, and so forth. Fill in the leaves for the following stemplot.

```
1|
1|
2|
2|
2|
2|
```

Give numerical summaries (center and spread) for this distribution and describe its main features (symmetry, skewness, outliers).

b) You can draw the boxplots by hand, or you can use computer software that will draw the plots once you have entered the data for the two groups. If you are drawing them by hand, first find the five-number summary for the midsize cars and for the four-wheel drive sports utility vehicles. The boxplot is then just a graphic that uses these five numbers.

Complete the five-number summary for both types of cars in the following table. It's simplest if you first list the miles per gallon in increasing order for each of the two groups. There are many "ties" in the data, so be careful when following the rules for finding quartiles.

Five-number summary	Midsize cars	Four-wheel drive SUVs
Minimum		
First quartile		
Median		
Third quartile		
Maximum		

Complete the boxplots in the space provided. Drawing these boxplots requires only the information in the five-number summaries.

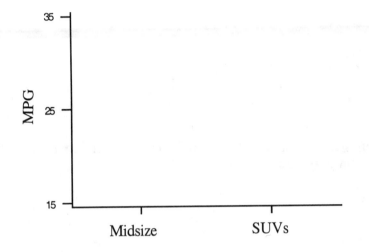

Exercise 1.81

KEY CONCEPTS - stemplots, five-number summary, describing a distribution

The stemplot produced by the software and provided in the problem has a format that is slightly different than other formats given in the text. If there are outliers, rather than producing the many extra stems required to include the outliers in the stemplot, the computer package lists the high and low outliers separately and does not include them in the stemplot. The four low outlying observations are -34.042255, -31.25, -27.06271, and -26.6129. The next three smallest observations are given in the first stem and are -19, -18 and -15. Similarly, at the upper end, there are five high outliers that are separated from the body of the stemplot. Summary statistics are also included in the computer output.

a) All numbers for the five-number summary can be read directly from the output.

b) Identify the rate of return for the best month. What would $1000 worth of Wal-Mart stock have been worth at the end of this month?

Now do the same for the worst month.

c) What are the main features of this distribution?

Exercise 1.83

KEY CONCEPTS - timeplot

Some of the interesting features in these data are the patterns that might occur over time. However, these features are lost when the data is presented in a stemplot. The side-by-side boxplots for the 19 years of data allow you to look for interesting patterns occurring over time.

a) To answer this, look at the medians (centers of the boxes) and see if you notice any patterns or long-term trends.

b) The spread for a given year can be examined in two ways. The first is to consider the length of the box (from first to third quartile) as a measure of spread, and the second is to use the length from the minimum to the maximum as the measure of spread. Often the two give similar impressions. The disadvantage of the second is that it is completely determined by only two observations (the minimum and the maximum), which may give a misleading impression of the spread for the remainder of the data.

c) Examining the minima and the maxima will show you where most of the outliers are. See if you can determine where some of the remaining ones are by a closer inspection of the boxplots.

Exercise 1.89

KEY CONCEPTS - finding the value x corresponding to a given proportion or area under an arbitrary normal density curve

To solve this problem, we must use an approach that is the reverse of that used in exercises 1.67 and 1.69. Conceptually, we think of having standardized the problem. We then find the value z for the standard normal distribution that satisfies the stated condition, i.e., has the desired area. We next must "unstandardize" this z value by first multiplying by the standard deviation and then adding the mean to the result. This unstandardized value x is the desired result.

Example 1.18 in the text summarizes the steps needed for such a "backward" normal calculation. We follow these steps.

State the problem. We first want to find the head size x_1 with area 0.05 (corresponding to 5%) to the left under the normal curve with mean $\mu = 22.8$ inches and standard deviation $\sigma = 1.1$ inches. This will tell us the head size below which are the smallest 5% of head sizes. We also want to find the head size x_2 with area 0.05 (corresponding to 5%) to the right under the normal curve with mean $\mu = 22.8$ inches and standard deviation $\sigma = 1.1$ inches. This will tell us the head size above which are the largest 5% of head sizes.

The figure states this question in graphical form.

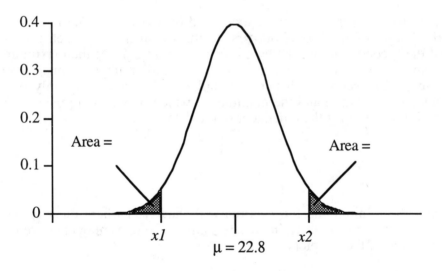

We need to express these areas as areas to the left of x_1 and x_2. What are these areas?

Use the table. Look in Table A for the entry closest to these two areas. For example, the area to the left of x_1 is 0.05. The entry in Table A with this area is about $z_1 = -1.65$. Thus $z_1 = -1.65$ is the standardized value with area 0.05 to its left. What is the area to the left of x_2 and what is the standardized value z_2 with this area to its left?

$z_2 =$

Unstandardize. To transform the solution from z_1 and z_2 back to the original x scale we use the "unstandardizing" formula

$$x = \mu + z\sigma$$

Applying this formula to z_1 yields

$x_1 = 22.8 + (z_1 \times 1.1) = 22.8 + (-1.65 \times 1.1) = 22.8 - 1.815 = 20.985$ inches

Now you find x_2.

$x_2 =$

State your conclusions.

COMPLETE SOLUTIONS

Exercise 1.73

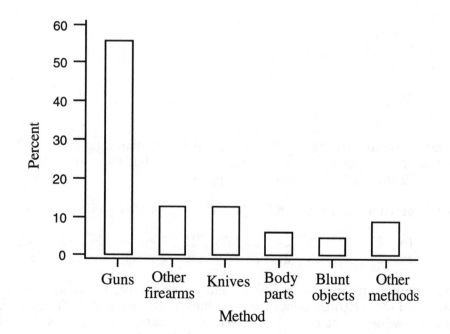

The "Other methods" category is needed if you do a bar chart or a pie chart since 8.8% of the murders are not included in the first five categories. The pie chart for these data follows.

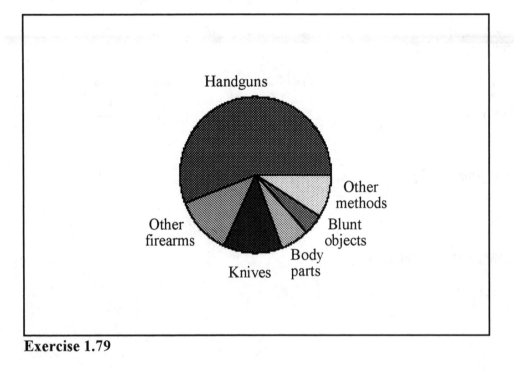

Exercise 1.79

a)
```
1 | 67
1 | 8899999999
2 | 0001
2 | 2
2 |
2 | 66
```

The distribution is reasonably symmetric with a median mpg of 19. There are two outliers corresponding to an mpg of 26. All but these two have an mpg between 16 and 22, and eight have an mpg of 19.

b) The ordered miles per gallon for the midsize cars are as follows:

16	23	23	24	24	25	<u>25</u>	25	25	26	26	26	**26**
26	27	28	28	28	28	<u>29</u>	29	29	29	30	30	33

The minimum and maximum are 16 and 33 respectively. When finding the median and the quartiles, concentrate on their positions, and don't pay attention to the tied values in the data. Since there are 26 cars, the median is the average of the 13th and 14th smallest values, which have been entered in bold-face.

Since these are both 26, the median is 26. There are 13 cars below and 13 cars above the average of the 13th and 14th. The first quartile is the median of the 13 observations below the median's position, which corresponds to the 7th smallest observation. It is underlined above. The third quartile is the 7th smallest of the 13 observations in the second row (those above the median position) and it is underlined as well.

The ordered miles per gallon for the four-wheel drive SUVs are as follows:

16   17   18   18   <u>19</u>   19   19   19   19   **19**   19   19   20
20   <u>20</u>   21   22   26   26

Since there are 19 data values, the median is the 10th smallest and is in bold. The first quartile is the median of the nine observations below the overall median and is underlined, and the third quartile is the median of the nine observations above and is also underlined.

Five number summary	Midsize cars	Four-wheel drive SUVs
Minimum	16	16
First Quartile	25	19
Median	26	19
Third Quartile	29	20
Maximum	33	26

The boxplot for the midsize cars is straightforward. The reason the boxplot looks unusual for the SUVs is because the first quartile and the median have the same value, so two of the lines in the box are in the same place. It's clear from the comparison of the boxplots that the SUVs tend to get lower mileage, and there is much less spread in the distribution of mileage for the SUVs.

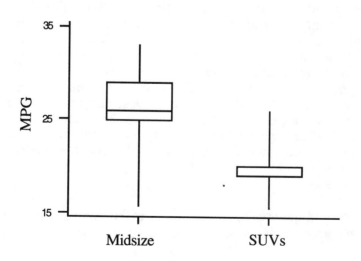

Exercise 1.81

a) The minimum is -34.04255 (the smallest low outlier) and the maximum is 58.67769 (the largest high outlier). The first quartile, median and third quartile are all listed as part of the computer output and are -2.950258, 3.4691 and 8.4511, respectively.

b) The highest rate of return is 58.67769%. A $1000 investment would have earned $1000 x 0.5867769 or $586.68 and have been worth $1586.68 at the end of the month. The lowest rate of return was -34.04255%. If you had invested this month you would have lost $340.43, so your initial investment of $1000 would have been worth $659.57 at the end of this month.

c) The distribution is fairly symmetric with a few high and low outliers. The mean and median are fairly close, which is a consequence of the roughly symmetric distribution.

Exercise 1.83

a) The first two years have a median rate of return that is negative. From 1975 onwards the median rate of return was positive (except for 1984) but fluctuated up and down without showing any long-term trend to either increase or decrease.

b) Looking at either the lengths of the boxes or the distance from the minimum to the maximum, the first three years have much larger spreads than succeeding years. The years from 1976 through 1991 have fairly similar spreads, the only exception being 1987 which has one low outlier.

c) Many of the outliers show up as either the maximum or minimum for a given year. The 58.7 occurred in 1973, and the 57.9 occurred in 1975. Since 1973 - 1975 are the only years with rates of return over 40, the 41.8 and the 42.0 must have occurred in these three years. One of these two values must be the maximum in 1974, although it is not possible to tell which from the graph. The 32 occurred in 1979 (although the value of the maximum in 1979 can't be determined exactly from the graph, the next largest observation is 24 and the maximum in 1979 clearly exceeds this value). The only years with rates of return below -20 were 1973 and 1987. The two lowest outliers, -34 and -31 came from 1973 and at least one of the outliers -27.1 and -26.6 came from 1987, the other possibly from 1973. These observations agree with the conclusion from part b, namely that the greatest spread occurred in the first three years.

Exercise 1.89

From the figure we see that the area to the left of x_1 is 0.05 and the area to the left of x_2 is 0.95. The corresponding standardized values are

$$z_1 = -1.65$$

$$z_2 = +1.65$$

The unstandardized values are

$x_1 = 22.8 + (z_1 \times 1.1) = 22.8 + (-1.65 \times 1.1) = 22.8 - 1.815 = 20.985$ inches

and

$x_2 = 22.8 + (z_2 \times 1.1) = 22.8 + (1.65 \times 1.1) = 22.8 + 1.815 = 24.615$ inches

We conclude that soldiers in the smallest 5% have head sizes below 20.985 inches and those in the largest 5% have head sizes above 24.615 inches. Thus soldiers with head sizes below 20.985 or above 24.615 inches get custom-made helmets.

CHAPTER 2

EXAMINING RELATIONSHIPS

SECTION 2.1

OVERVIEW

Chapter 1 provides the tools to explore several types of variables one by one, but in most instances the data of interest are a collection of variables that may exhibit relationships among themselves. Typically, these relationships are more interesting than the behavior of the variables individually. The first tool we consider for examining the relationship between variables is the **scatterplot**. Scatterplots show us two quantitative variables at a time, such as the weight of a car and its miles per gallon (MPG). Using colors or different symbols, we can add information to the plot about a third variable that is categorical in nature. For example, if in our plot we wanted to distinguish between cars with manual or automatic transmissions, we might use a circle to plot the cars with manual transmissions and a cross to plot the cars with automatic transmissions.

When drawing a scatterplot, we need to pick one variable to be on the horizontal axis and the other to be on the vertical axis. When there is a **response variable** and an **explanatory variable**, the explanatory variable is always placed on the horizontal axis. In cases where there is no explanatory-response variable distinction, either variable can go on the horizontal axis. After drawing the scatterplot by hand or using a computer, the scatterplot should be examined for an **overall pattern** that may tell us about any relationship between the variables and for **deviations** from it. You should be looking for the **direction**, **form**, and **strength** of the overall pattern. In terms of direction, **positive association** occurs when the variables both take on high values together, while **negative association** occurs if one variable takes high values when the other takes on low values. In many cases, when an association is present, the variables appear to have a **linear relationship**. The plotted values seem to form a line. If the line slopes up to the right, the association is positive; if the line slopes down to the right, the association is negative. As always, look for **outliers**. The outlier may be far away in terms of the horizontal variable or the vertical variable or far away from the overall pattern of the relationship.

GUIDED SOLUTIONS

Exercise 2.3

KEY CONCEPTS - explanatory and response variables, categorical and quantitative variables

When examining the relationship between two variables, if you hope to show that one of the variables can be used to explain variation in the other, remember that the **response variable** measures the outcome of the study, while the **explanatory variable** explains or causes changes in the response variable. When you just want to explore the relationship between two variables, such as score on the math and verbal SAT, then the explanatory-response variable distinction is not important.

If you do not recall the distinction between a categorical and a quantitative variable, refer to the Introduction to Chapter 1.

In this problem, the response variable is how long a patient lived after treatment, since it is the outcome of the study. Is this variable categorical or quantitative?

Now it is your turn.

　　　　explanatory variable =

Is this variable categorical or quantitative?

Exercise 2.7

KEY CONCEPTS - drawing and interpreting a scatterplot, adding a categorical variable to a scatterplot

a) When drawing a scatterplot, we first need to pick one variable (the explanatory variable) to be on the horizontal axis and the other (the response) to be on the vertical axis. In this data set we are interested in the "effect" of lean body mass on metabolic rate. So lean body mass is the explanatory variable and metabolic rate is the response variable in the plot in the following figure. Although you will generally draw scatterplots on the computer, drawing a small one like this by hand makes sure that you understand what the points represent.

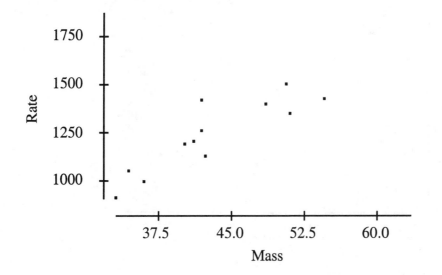

b) Here are some guidelines for examining scatterplots: Do the data show any association? **Positive association** is when the variables both take on high values together. **Negative association** is when one variable takes high values and the other takes on low values. If the plotted values seem to form a line, the variables may have a **linear relationship**. If the line slopes up to the right, the association is positive. If the line seems to slope down to the right, the association is negative. Are there any **clusters** of data? Clusters are distinct groups of observations. As always, look for outliers. An **outlier** may be far away in terms of the horizontal variable or the vertical variable, or far away from the overall pattern of the relationship.

For our plot, is the association positive or negative? Do females with higher lean body mass tend to have higher or lower metabolic rates?

What is the form of the relationship? Do the points in the plot tend to follow a straight-line pattern? A curved pattern? Are distinct clusters present?

How strong is the relationship?

c) The scatterplot for the females is repeated here. Add the data for the males to this plot using a different color or plotting symbol.

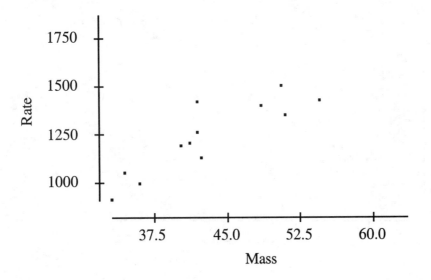

Is the pattern of the relationship for the men similar to that for the female subjects? If not, how does it differ?

How do the male subjects as a group differ from the female subjects as a group?

Exercise 2.11

KEY CONCEPTS - drawing and interpreting a scatterplot

a) When drawing a scatterplot, we first need to pick one variable (the explanatory variable) to be on the horizontal axis and the other (the response) to be on the vertical axis. In this data set we are interested in the "effect" of drinking moderate amounts of wine on yearly deaths from heart disease. So wine consumption is the explanatory variable and deaths from heart disease is the response. We have drawn the points corresponding to Australia and Austria in the plot below. Although in practice people often draw scatterplots on the computer, drawing a small one like this by hand makes sure that you understand what the points represent.

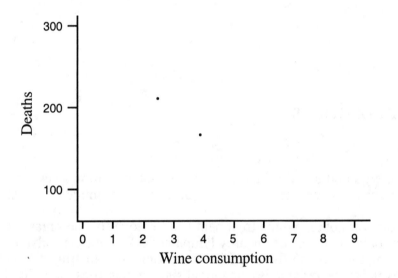

b) We are looking for the form and strength of the relationship. Can the relationship be described with a straight line? Section 2.3 discusses formal methods for drawing a straight line through a set of data, but for now just try to draw a straight line to follow the pattern in the scatterplot in part a. Do the points seem to follow the line you have drawn? How tight is the scatter about the line?

c) Is the association positive or negative? Do countries with higher wine consumption tend to have higher or lower death rates? Be careful with the language you use to describe the relationship. In this example, the countries may differ on many other factors besides wine consumption, which may explain the lower death rates due to heart attacks. So avoid the use of expressions such as "drinking more wine lowers the risk of heart disease" or "wine produces a lower risk of heart disease," as these expressions imply causation. Try and explain in simple language what the data are saying.

COMPLETE SOLUTIONS

Exercise 2.3

As stated in the guided solution, the response variable is how long a patient lived after treatment. Since it is a measure of time, it is a quantitative variable.

The explanatory variable is which treatment (the removal of the breast versus the removal of only the tumor and nearby lymph nodes) a patient received. The treatments attempt to explain the response (how long after treatment a patient lived). The explanatory variable is categorical since it classifies the individuals into one of two treatment categories.

Exercise 2.7

a) The graph given in the Guided Solution.

b) As lean body mass increases, or as you move from left to right across the horizontal axis in the scatterplot, the points in the plot tend to rise. This indicates that the association between the variables is positive. The form of the relationship appears to be linear since a straight line seems to be a reasonable approximation to the overall trend in the plot. The relationship is not perfect, but it appears to be moderately strong.

c) We add the men to the plot. Men are indicated by the x's.

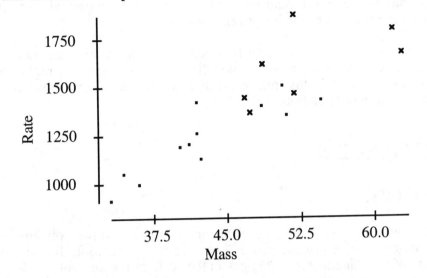

The male subjects (the x's) also show a positive association that might be described as linear. The association does not appear so strong as for the women and the slope of the linear relation may be a bit flatter. We also notice that the men are clustered in the upper right of the plot. This is not surprising, since men tend to be larger than women.

Exercise 2.11

a, b)

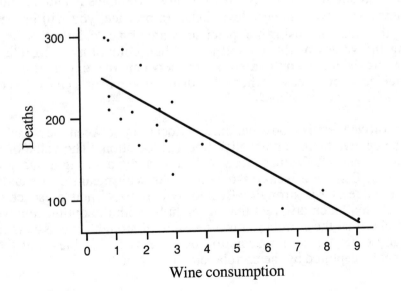

A line does not do a bad job of describing the general pattern. The relationship is moderately strong. In Section 2.2 we will give a numerical measure that describes the strength of the linear relationship.

c) There is a negative association between wine consumption and deaths from heart disease. Those countries in which wine consumption is higher tend to have a lower rate of deaths from heart disease. (Note: There is nothing in this language that implies causation.)

SECTION 2.2

OVERVIEW

Scatterplots provide a visual tool for looking at the relationship between two variables. Unfortunately, our eyes are not good tools for judging the strength of the relationship. Changes in the scale or the amount of white space in the graph can easily change our judgment of the strength of the relationship. **Correlation** is a numerical measure we use to show the strength of **linear association**.

The correlation can be calculated using the formula

$$r = \frac{1}{n-1} \sum (\frac{x_i - \bar{x}}{s_x})(\frac{y_i - \bar{y}}{s_y})$$

where $\bar{x}$ and $\bar{y}$ are the respective means for the two variables X and Y, and s_x and s_y are their respective standard deviations. In practice, you will probably be computing the value of r using computer software or a calculator that finds r from entering the values of the x's and y's. When computing a correlation coefficient, there is no need to distinguish between the explanatory and response variables, even in cases where this distinction exists. The value of r does not change if we switch x and y.

When r is is positive there is a positive linear association between the variables, and when r is negative there is a negative linear association. The value of r is always between 1 and -1. Values close to 1 or -1 show a strong association, while values near 0 show a weak association. As with means and standard deviations, the value of r is strongly affected by outliers. Their presence can make the correlation much different than it might be with the outlier removed. Finally, remember that the correlation is a measure of straight line association. There are many other types of association between two variables, but these patterns will not be captured by the correlation coefficient.

GUIDED SOLUTIONS

Exercise 2.17

KEY CONCEPTS - scatterplots and computing the correlation coefficient

a) For these data we do not envision one of the variables as explaining the other. All we are interested in is investigating the association between the two variables. We are free to arbitrarily designate one of the variables as the explanatory variable and the other as the response. We choose the variable Femur as the explanatory variable, plotting it on the horizontal axis, and choose Humerus to play the role of the response, plotting it on the vertical axis. Use the axes provided to make your scatterplot.

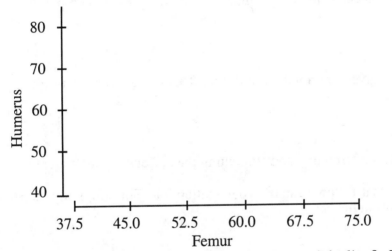

Do any of the points appear to depart from a straight line? What does this imply about whether all five specimens came from the same species?

b) Let x denote the Femur measurements and y the Humerus measurements. We find the means and standard deviations to be

$$\bar{x} = 58.20, \quad s_x = 13.1985$$

$$\bar{y} = 66.00, \quad s_y = 15.8902$$

Calculations by hand are best done systematically, such as in the following table. The second and fourth columns are the standardized values for x and y. We have provided the table entries for the first two x, y values. See if you can complete the remaining entries

x	$\left(\dfrac{x-\bar{x}}{s_x}\right)$	y	$\left(\dfrac{y-\bar{y}}{s_y}\right)$	$\left(\dfrac{x-\bar{x}}{s_x}\right)\left(\dfrac{y-\bar{y}}{s_y}\right)$
38	-1.5305	41	-1.5733	2.4079
56	-0.1667	63	-0.1888	0.0315
59		70		
64		72		
74		84		

Now sum up the values in the last column and divide by n-1 to compute r.

$r =$

c) Our calculator gives a correlation of 0.994. This agrees with the result in part a.

Exercise 2.22

KEY CONCEPTS - interpreting and computing the correlation coefficient

a) Use the axes that follow to make your scatterplot. For simplicity, consider using the symbols in the key.

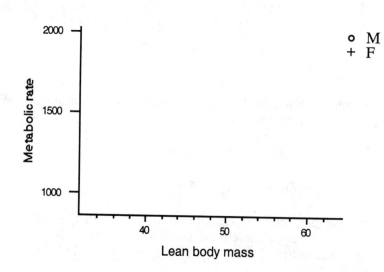

Should the sign of the correlation coefficient be the same for men and women? Is either relationship "stronger"? Are there outliers in either group that might raise or lower the value of the correlation coefficient?

b) Use your calculator (or computer if you are using statistical software) to compute the correlation coefficient. Record your results.

Men's correlation coefficient =

Women's correlation coefficient =

c) This simply involves computing means. Record your results below.

Mean body mass for men =

Mean body mass for women =

This is a difficult question. If the relationship remained the same for all weights between 40 and 65 kilograms (the range of all the data), then it wouldn't matter. The fact that the men were heavier wouldn't influence the correlation. If the relationship was different for heavier people than lighter people, then the fact that men are heavier could affect the value of the correlation, since then men and women might have a different relationship between body mass and metabolic rate (with possibly differing strengths as measured by the correlation coefficient).

d) Is this a linear transformation? What is the effect of a linear transformation on the correlation coefficient?

Exercise 2.25

KEY CONCEPTS - changing units of measurement

a) Use the axes given for your plot.

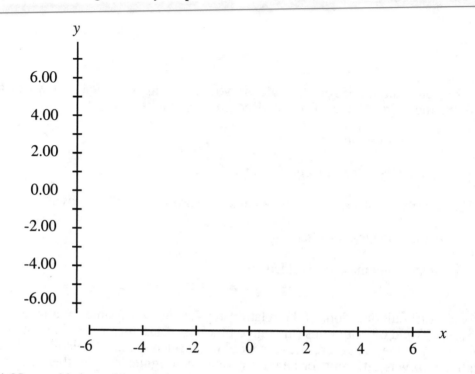

b) Now add the values of the new variables to your plot. Remember to use a different plotting symbol.

c) Use your calculator (or computer software) to calculate the correlations. What do you observe about their values? Why should you have expected this?

Exercise 2.27

KEY CONCEPTS - interpreting the correlation coefficient

a) The key phrase is "A well-diversified portfolio includes assets with low correlations." What does this imply about which of the two investments she should choose?

b) What can you say about the correlation between two variables when increases in one of the variables are associated with decreases in the other?

COMPLETE SOLUTIONS

Exercise 2.17

a) The scatterplot follows.

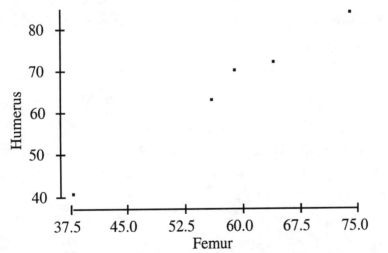

One of the points (the third point, with Femur = 59 and Humerus = 70) appears to differ a bit from the others. If this point is ignored, the other four appear to lie nearly exactly on a straight line. While the difference does not appear to be dramatic, this point might come from a different species than the others. The evidence does not appear overwhelming, however.

b) The completed table follows.

x	$\left(\dfrac{x - \bar{x}}{s_x}\right)$	y	$\left(\dfrac{y - \bar{y}}{s_y}\right)$	$\left(\dfrac{x - \bar{x}}{s_x}\right)\left(\dfrac{y - \bar{y}}{s_y}\right)$
38	-1.5305	41	-1.5733	2.4079
56	-0.1667	63	-0.1888	0.0315
59	0.0606	70	0.2517	0.0153
64	0.4394	72	0.3776	0.1659
74	1.1971	84	1.1328	1.3561

The sum of the values in the last column is 3.9767. Thus the correlation is

$$r = 3.9767/4 = 0.9942.$$

c) Our calculator gives a correlation of 0.994. This agrees with the result in part a.

Exercise 2.22

a) The scatterplot follows.

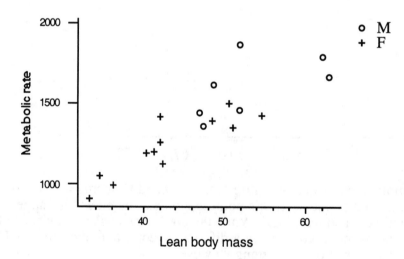

The relationship for the women seems to be tighter around a line. The men's observation of mass = 51.9 and rate = 1867 lowers the value of the correlation for the men.

b) You should try to use a computer package or a calculator to compute the value of the correlation coefficient. If you do not have access to a calculator or computer package, or if you do not trust your calculations, the required "hand" computations are given below for the men. You can use them to double check your work if you made an error.

$$\bar{x} = 53.10, \quad s_x = 6.69$$

$$\bar{y} = 1600.00, \quad s_y = 189.2$$

We summarize the calculations for the correlation r in the following table

x	$\left(\dfrac{x - \bar{x}}{s_x}\right)$	y	$\left(\dfrac{y - \bar{y}}{s_y}\right)$	$\left(\dfrac{x - \bar{x}}{s_x}\right)\left(\dfrac{y - \bar{y}}{s_y}\right)$
62.0	1.33034	1792	1.01480	1.35003
62.9	1.46487	1666	0.34884	0.51100
47.4	-0.85202	1362	-1.25793	1.07178
48.7	-0.65770	1614	0.07400	-0.04867
51.9	-0.17937	1460	-0.73996	0.13273
51.9	-0.17937	1867	1.41121	-0.25313
46.9	-0.92676	1439	-0.85095	0.78862

The sum of the values in the last column is 3.5524. Thus the correlation is

$$r = 3.5524/6 = 0.592$$

for the men. Thus,

Men's correlation coefficient = 0.592

Likewise for the women we find

Women's correlation coefficient = 0.876

c) In part b above we see

Mean body mass for men = 53.10.
Mean body mass for women = 43.03

The rest of the question is answered in the Guided Solution.

d) Kilograms = 2.2 x pounds is a linear transformation with $a = 0$ and $b = 2.2$ (see page 56 of the text). The value of the correlation is unchanged under linear transformations of the variables.

Exercise 2.25

a) Here is a completed scatterplot of the data.

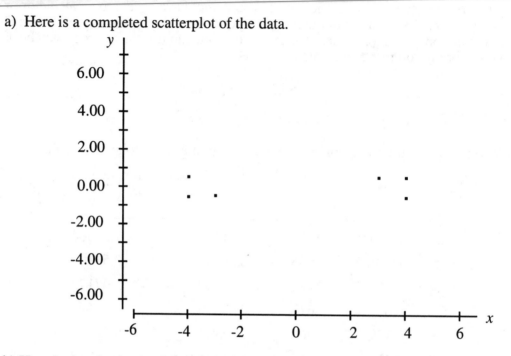

b) Here is the plot in a) with the x^* and y^* points added as the plotting symbol x.

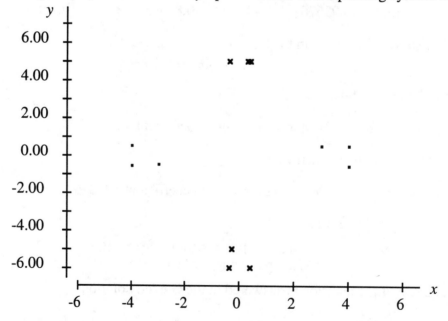

c) The correlation between x and y is 0.253 (as calculated by our calculator). The correlation between x^* and y^* is also 0.253. The two correlations are the same. This is not surprising since x^* is a multiple (1/10) of x and y^* is a multiple (10) of y. This means that both x^* and x will have the same standardized values. Likewise, y^* and y will also have the same standardized values. As a result, the correlations between x and y and between x^* and y^* will be the same.

Exercise 2.27

a) Since Rachel should be looking for investments that have a weak correlation with municipal bonds, she should invest in the small-cap stocks. The correlation between municipal bonds and the small-cap stock is weaker (closer to 0) than the correlation between municipal bonds and the large-cap stocks.

b) When increases in one variable are associated with decreases in another, the two variables are negatively associated and the correlation is negative. Thus Rachel should look for assets that are negatively correlated with municipal bonds. Neither large-cap nor small-cap stocks meet this criterion.

SECTION 2.3

OVERVIEW

If a scatterplot shows a linear relationship that is moderately strong as measured by the correlation, we can draw a line on the scatterplot to summarize the relationship. In the case where there is a response and an explanatory variable, the **least-squares regression** line often provides a good summary of this relationship. A straight line relating y to x has the form $y = a + bx$, where b is the **slope** of the line and a is the **intercept**. The least-squares regression line is the straight line $\hat{y} = a + bx$, which minimizes the sum of the squares of the vertical distances between the line and the observed values y. The formula for the slope of the least squares line is

$$b = r\frac{s_y}{s_x}$$

and for the intercept is $a = \bar{y} - b\bar{x}$, where $\bar{x}$ and $\bar{y}$ are the means of the x and y variables, s_x and s_y are their respective standard deviations, and r is the value of the correlation coefficient. Typically, the equation of the least-squares regression line is obtained by computer software or a calculator with a regression function.

Regression can be used to predict the value of y for any value of x. Just substitute the value of x into the equation of the least-squares regression line to get the predicted value for y.

Correlation and regression are clearly related, as can be seen from the equation for the slope, b. However, the more important connection is how r^2, the square of the correlation coefficient, measures the strength of the regression. r^2 tells us the fraction of the variation in y that is explained by the regression of y on x. The closer r^2 is to 1, the better the regression describes the connection between x and y.

An examination of the **residuals** shows us how well our regression does in predictions. The difference between an observed value of y and the predicted value obtained by least-squares regression, $\hat{y}$, is called the residual.

$$\text{residual} = y - \hat{y}$$

Plotting the residuals is a good way to check the fit of our least-squares regression line. Features to look for in a **residual plot** are unusually large values of the residuals (outliers), nonlinear patterns, and uneven variation about the horizontal line through zero (corresponding to uneven variation about the regression line).

Also look for influential observations. **Influential observations** are individual points whose removal would cause a substantial change in the regression line. Influential observations are often outliers in the horizontal direction.

GUIDED SOLUTIONS

Exercise 2.31

KEY CONCEPTS - drawing and interpreting the least-squares regression line

a) Perhaps the simplest way to draw a graph of the line given is to pick two convenient values for the variable weeks, substitute them into the equation of the least-squares regression line, and compute the corresponding value of pH predicted by the equation for each. This produces two sets of weeks and pH values. Each of these week and pH pairs corresponds to a point on the least-squares regression line. Simply plot these points on your graph and connect them with a straight line. Be sure that the horizontal axis of your graph corresponds to the explanatory variable or x value in the equation of the line (in this case, weeks) and the vertical axis to the response variable or y value in the equation (in this case, pH)

Convenient values for weeks might be 0 and 100. Complete the following.

For weeks = 0, pH =

For weeks = 100, pH =

Now plot these two sets of week and pH values as points on the axes that follow. Then connect them with a straight line.

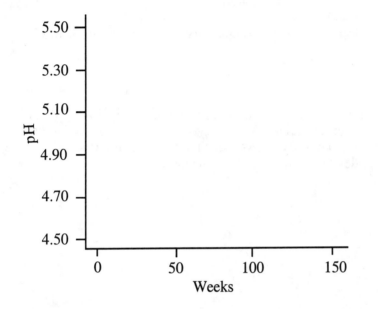

From the least-squares regression line, does pH increase or decrease as weeks increases? Is the association, therefore, positive or negative?

Write in plain language what this association means.

b) According to the equation of the least-squares regression line, namely,

$$pH = 5.43 - (0.0053 \times weeks)$$

what is the value of pH at the beginning and end of the study?

For weeks = 1, pH =

For weeks = 150, pH =

c) For a line whose equation is

$$y = a + bx$$

the quantity b is the slope. Identify this quantity in the equation of the least-squares regression line.

Slope =

Explain clearly in writing what this slope says about the change in the pH of the precipitation in this wilderness area. (Writing your explanation out forces you to express your thoughts explicitly. If you can't write a clear explanation, you may not adequately understand what the slope means.)

Exercise 2.33

KEY CONCEPTS - least-squares regression, prediction

a) Using the same software that generated Figure 2.12c, we obtain the following output.

Dependent variable is: **Pulse**
No Selector
R squared = 55.6%   R squared (adjusted) = 53.5%
s = 6.455 with 23 - 2 = 21 degrees of freedom

Source	Sum of Squares	df	Mean Square	F-ratio
Regression	1097.94	1	1097.94	26.3
Residual	875.021	21	41.6677	

Variable	Coefficient	s.e. of Coeff	t-ratio	prob
Constant	479.934	66.23	7.25	0.0001
Time	-9.69490	1.889	-5.13	0.0001

The intercept can be found in the column headed "Coefficient" in the next to last line of the output, the slope in the same column, last line. After rounding, we verify that the equation of the least-squares regression line is

$$Pulse = 479.9 - 9.695 \ time$$

Depending on the software you used, your output may look a bit different, but you should obtain (after rounding) the same equation for the least-squares regression line. Similarly, if you used a calculator that allows you to compute the least-squares regression line, you should (after rounding) obtain the same equation.

If your results don't agree, you may have made an error in the data entry. Double check that you have entered the data correctly, since this is perhaps the most common mistake one can make.

b) To predict the professor's pulse rate, substitute the time of 34.30 minutes into the equation of the least-squares regression line and compute the pulse rate.

$$Pulse = 479.9 - 9.695 \times 34.30 =$$

What is the difference between the value you computed and the actual pulse rate of 152?

c) Notice that we now want to predict time from pulse. Which variable is now the predictor and which the response? How does this affect the computation of the least-squares regression line? Use your calculator or software to compute this new least-squares regression line and write the equation.

Based on this new equation, what do you predict the time to be?

$$Time =$$

How accurate is this prediction?

d) Write your explanation in the space provided. Remember, you are writing to someone who knows no statistics, so you shouldn't use statistical jargon.

Exercise 2.37

KEY CONCEPTS - outliers and influential observations

a) Draw your scatterplot on the axes that follow. Remember to circle the points for spaghetti and snack cake on your plot. Note that since we wish to predict guessed calories from true calories, true calories is the predictor variable, guessed calories the response.

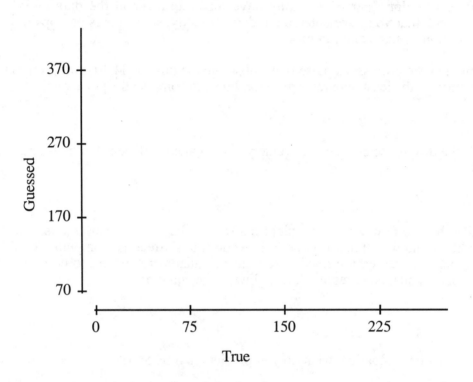

b) Use your calculator or statistical software to compute the equations of the least-squares regression lines. Record your results.

Equation of least-squares regression line for all ten points:

Equation of least-squares regression line after leaving out spaghetti and snack cake:

c) Sketch these two lines into your plot in part a. To help you distinguish them, make one of them dashed. To decide if spaghetti and snack cake taken together are influential, ask yourself if their removal resulted in a marked change in the equation of the least-squares regression line. You may wish to refer to Figure 2.17 in your text to give you a sense of how much change might be considered marked.

Exercise 2.41

KEY CONCEPTS - calculating the least-squares regression line from summary statistics, r and r^2, prediction

a) Recall that if the least-squares regression line has the equation

$$\hat{y} = a + bx$$

the formula for the slope of the least-squares line is

$$b = r\frac{s_y}{s_x}$$

and for the intercept is

$$a = \bar{y} - b\bar{x}$$

where $\bar{x}$ and $\bar{y}$ are the means of the x and y variables, s_x and s_y are their respective standard deviations, and r is the value of the correlation coefficient.

Now use the values for $\bar{x}$, $\bar{y}$, s_x, s_y and r given in the problem to compute the slope b and intercept a.

$b =$ $\qquad\qquad\qquad$ $a =$

Equation of least-squares regression line: $\hat{y} =$

b) What quantity tells you the fraction of the variation in the values of y (in this case GPA) that is explained by the least-squares regression of y (or GPA) on x (in this case IQ)? Compute this quantity. Remember to convert the fraction to a percent.

c) In part a you should have found that the equation of the least-squares regression line is

$$GPA = -3.552 + 0.101 \times IQ$$

Use this equation to predict the value of GPA for a student with an IQ of 103.

Predicted GPA =

What, therefore, is the residual for the student with an IQ of 103 and a GPA of 0.53?

Residual = observed y - predicted y =

Exercise 2.45

KEY CONCEPTS - scatterplots, r and r^2, the least-squares regression line from summary statistics, prediction, residuals, influential observations

Note: Because of the size of this data set, this problem is best done with a calculator or, better yet, with statistical software.

a) Use the following axes below for your plot. Be sure to label the axes correctly (which variable is the predictor and which is the response?)

b) Give in the values of the correlation r and r^2.

$r =$

$r^2 =$

Now describe the relationship between U.S. and overseas returns in words. Use r and r^2 to make your description more precise. You may find it helpful to refer to the facts about correlation in Section 2 of this chapter and to Fact 4 in this section.

c) You will want to use your calculator or statistical software to compute the equation of the least-squares regression line. Write the equation.

$\hat{y} =$

Now draw this line in your plot in part a.

d) Use the equation of the least-squares regression line you found in part c to predict the return on overseas stock when the return on U.S. stocks is 33.4%.

Prediction =

How does your prediction compare to the 1997 overseas return of 2.1% when the U.S. return was 33.4%? Are you confident that predictions using the regression line will be quite accurate? Why? You may wish to think about what the answer to part b implies about prediction.

e) Circle the point in your plot in part a that has the largest residual. This point corresponds to what year (refer to Table 2.7 to answer this)?

Year =

Are there any points in your plot that seem likely to be very influential? Which one(s)?

COMPLETE SOLUTIONS

Exercise 2.31

a) We find

$$\text{For weeks} = 0, \text{pH} = 5.43 - (0.0053 \times 0) = 5.43$$

$$\text{For weeks} = 100, \text{pH} = 5.43 - (0.0053 \times 100) = 4.90$$

These two points are plotted below (using x as the plotting symbol) and the least-squares regression line drawn connecting them.

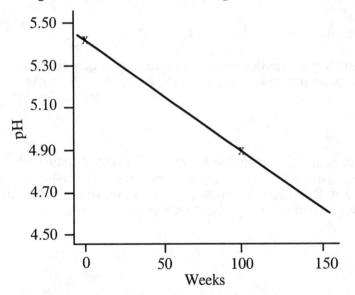

The line has a negative slope: as the variable weeks increases, pH decreases. This implies that the association is negative. In plain language, this association means that over the period of the study, as time passed the pH of the precipitation tended to decrease and hence the acidity of the precipitation tended to increase.

b) From the equation of the least-squares regression line, namely,

$$\text{pH} = 5.43 - (0.0053 \times \text{weeks})$$

we find

$$\text{For weeks} = 1, \text{pH} = 5.43 - (0.0053 \times 1) = 5.4247$$

$$\text{For weeks} = 150, \text{pH} = 5.43 - (0.0053 \times 150) = 4.6350$$

c) From the equation of the least-squares regression line, namely,

$$\text{pH} = 5.43 - (0.0053 \times \text{weeks})$$

we see that the slope is -0.0053. The slope tells us that, on average, for each week of the study period, the pH of the precipitation in this wilderness area decreased by 0.0053.

Exercise 2.33

a) The Guided Solution is a complete solution.

b) The predicted pulse rate is

$$\text{Pulse} = 479.9 - 9.695 \times 34.30 = 479.9 - 332.5385 = 147.3615$$

We round this off to 147.4.

The difference between this value and the actual pulse rate of 152 is (ignoring signs) 4.6. While not exactly correct, this is moderately accurate.

c) Pulse is now the predictor variable and time the response. We must recompute the least-squares regression line using pulse as the predictor and time as the response.

Dependent variable is: **Time**
No Selector
R squared = 55.6%   R squared (adjusted) = 53.5%
s = 0.4967 with 23 - 2 = 21 degrees of freedom

Source	Sum of Squares	df	Mean Square	F-ratio
Regression	6.50053	1	6.50053	26.3
Residual	5.18073	21	0.246701	

Variable	Coefficient	s.e. of Coeff	t-ratio	prob
Constant	43.0973	1.569	27.5	0.0001
Pulse	-0.057401	0.0112	-5.13	0.0001

The equation of the new least-squares regression line is (after rounding)

$$\text{Time} = 43.10 - (0.057 \times \text{pulse})$$

Based on this equation, our prediction is

$$\text{Time} = 43.10 - (0.057 \times 152) = 43.10 - 8.664 = 34.436$$

which we round off to 34.44. Compared to the actual time of 34.30 minutes, this isn't too bad a prediction (we are about 8 seconds too high).

d) In part a the line is appropriate for predicting pulse from time. In part c the line is appropriate for predicting time from pulse. The lines differ because the variable being predicted and the variable used to make the prediction are interchanged.

Exercise 2.37

a) Here is the scatterplot with the points for spaghetti and snack cake circled.

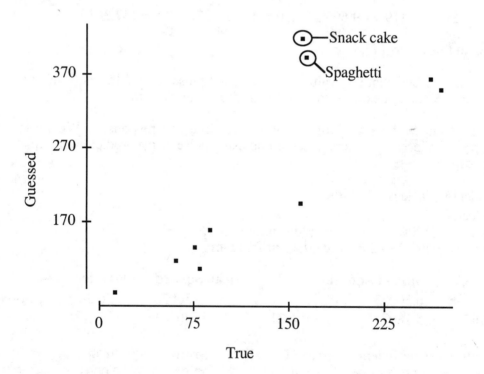

b) Here are the equations of the least-squares regression lines. You should obtain similar results using a calculator or statistical software, possibly after rounding.

For all ten data points:

$$\text{Guessed calories} = 58.588 + 1.304 \text{ x (true calories)}$$

After leaving out spaghetti and snack cake:

$$\text{Guessed calories} = 43.881 + 1.147 \text{ x (true calories)}$$

c) Here is our plot containing both lines.

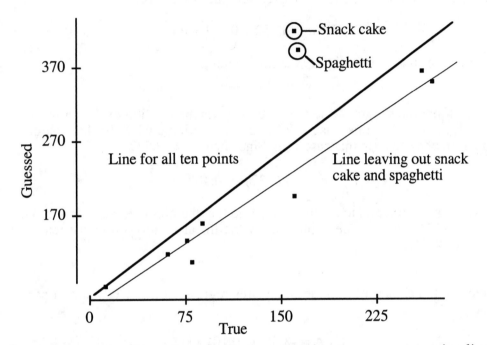

Removing snack cake and spaghetti causes the least-squares regression line to shift a noticeable amount (compare with Figure 2.17 in the text). We would consider spaghetti and snack cake taken together as influential.

Exercise 2.41

a) In the problem we are given that

$$\bar{x} = 108.9 \qquad s_x = 13.17$$

$$\bar{y} = 7.447 \qquad s_y = 2.10$$

$$r = 0.6337$$

Thus the slope is

$$b = r\frac{s_y}{s_x} = 0.6337\frac{2.10}{13.17} = 0.101$$

and the intercept is

$$a = \bar{y} - b\bar{x} = 7.447 - 0.101 \times 108.9 = 7.447 - 10.999 = -3.552$$

The equation of the least-squares regression line is therefore

$$GPA = -3.552 + 0.101 \times IQ$$

(Your answer may differ slightly due to rounding.)

b) Recall that the square of the correlation, r^2, is the fraction of the variation in the values of y (in this case GPA) that is explained by the least-squares regression of y on x (in this case IQ). Since here $r = 0.6337$,

$$r^2 = (0.6337)^2 = 0.402$$

Converting this fraction to a percent we find that the observed variation in these students' GPAs that can be explained by the linear relationship between GPA and IQ is 40.2%.

c) We use the equation of the least-squares regression line we found in part a, namely,

$$GPA = -3.552 + 0.101 \times IQ$$

to make our prediction. We calculate

$$\text{Predicted GPA} = -3.552 + 0.101 \times 103 = 6.851$$

and hence for an observed GPA of 0.53,

$$\text{Residual} = \text{observed } y - \text{predicted } y = 0.53 - 6.851 = -6.321$$

Exercise 2.45

a) In this problem, we used statistical software to create our plots and carry out computations. Our scatterplot follows.

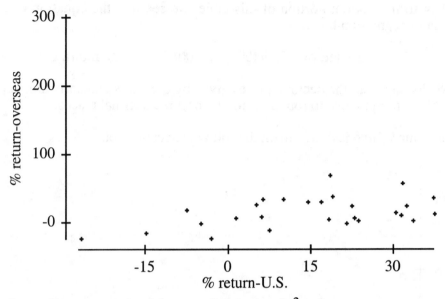

b) Our software computed the correlation r and r^2 to be

$$r = 0.464, \qquad r^2 = (0.464)^2 = 0.215$$

Since the correlation $r = 0.464$, there is a positive association between overseas and U.S. return. In other words, above-average U.S. returns tend to be associated with above-average overseas returns. Likewise, below-average U.S. returns tend to be associated with below-average overseas returns. The strength of the association is not strong. Since $r^2 = 0.215$, the fraction of the variation in overseas returns that is explained by the least-squares regression of overseas return on U.S. returns is only 21.5%.

c) We obtain the following table from our statistical software. This is the same software used to generate Figure 2.12c in the text. You may wish to refer to that figure to help you understand the table.

Dependent variable is: **%return-overseas**
No Selector
R squared = 21.5%   R squared (adjusted) = 18.4%
s = 19.87 with 27 - 2 = 25 degrees of freedom

Source	Sum of Squares	df	Mean Square	F-ratio
Regression	2708.99	1	2708.99	6.86
Residual	9866.85	25	394.674	

Variable	Coefficient	s.e. of Coeff	t-ratio	prob
Constant	5.69400	5.138	1.11	0.2783
%return-US	0.620082	0.2367	2.62	0.0147

According to the bottom portion of this table, we see that the equation of the least-squares regression line is

$$\% \text{ overseas return} = 5.69400 + 0.620082 \times \% \text{ U.S. return}$$

We have displayed all the decimal places given by our statistical software, but it is probably better practice to round off to two or three decimal places.

The least-squares line is drawn in on the following scatterplot.

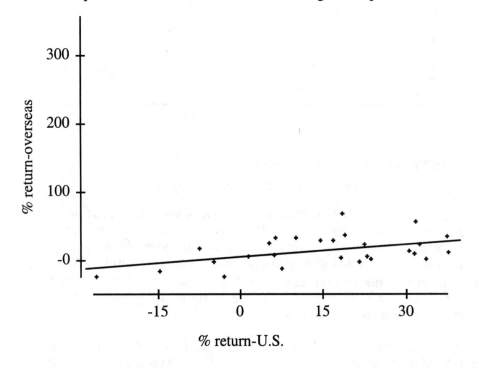

d) Using the equation of the least-squares regression line we obtained in part c, we predict

$$\% \text{ overseas return} = 5.69400 + 0.620082 \times 33.4\% = 26.40\%$$

This does not compare well with the overseas return of 2.1% in 1997 when U.S. returns were 33.4%. We should not be overly surprised at the poor quality of our predictions. Remember, $r^2 = 0.215$, so the fraction of the variation in overseas returns that is explained by the least-squares regression of overseas returns on U.S. returns is only 21.5%.

e) The point with the largest residual is circled in the plot below.

This corresponds to

$$Year = 1986.$$

None of the points in the plot appears likely to be very influential.

SECTION 2.4

OVERVIEW

Section 4 is the warning label that comes with regression tools. The first parts of this chapter present some new and potentially powerful ways of describing the relation between two variables. Now here are some limitations.

Watch out for the following:

- Do not **extrapolate** beyond the range of the data.

- Be aware of possible **lurking variables**.

- Are the data averages or from individuals? **Averaged data** usually lead to overestimating the correlations.

- Most of all, remember that *association is not causation!* Just because two variables are correlated doesn't mean one causes changes in the other.

GUIDED SOLUTIONS

Exercise 2.53

KEY CONCEPTS - scatterplots, least-squares regression, r^2, extrapolation

a) On the axes provided, draw the scatterplot. Which is the response variable and which the explanatory variable? Which goes on the vertical and which on the horizontal axis? Be sure to label axes clearly.

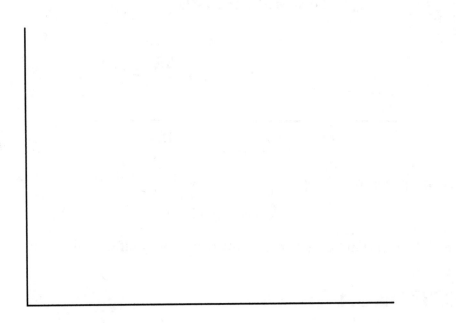

Use your calculator or statistical software to find the least-squares regression line. Write the equation.

b) What quantity in the least-squares regression line represents the average decline per year in farm population during this period? What is its value? In answering this question, remember to keep in mind the units of the variable Population.

What quantity represents the percent of the observed variation in farm population that is accounted for by linear change over time? What is the value of this quantity?

c) Use the equation of the least-squares regression line you computed in part a to answer this question. Again, remember the units of the variable Population in reporting your answer. Is your prediction reasonable? Why?

Exercise 2.55

KEY CONCEPTS - lurking variables

Try to think of variables that contribute both to heavy TV watching and to poor grades. To get you started, here is an example. A variable that reflects a child's attention span might be a lurking variable. Children with low attention spans may choose TV watching over reading or other activities requiring more concentration. Likewise, children with poor attention spans may not pay attention in class, leading to lower grades.

Now list some other variables (at least one). Be sure to explain how they contribute to heavy TV watching and poor grades.

Exercise 2.63

KEY CONCEPTS - r^2, correlations based on averages

a) Compute $r^2 = (0.970)^2 =$

b) How do correlations based on averages compare to those computed from the actual individuals? Why? If you are not sure how to answer this question, take a look at the Section 2.4 Summary in the text.

Exercise 2.65

KEY CONCEPTS - explanatory and response variables, lurking variable

Complete the following. Think about what sort of student takes two or more years of foreign language study.

Explanatory variable =

Response variable =

Lurking variable =

Why does this lurking variable prevent the conclusion that language study improves students' English scores?

COMPLETE SOLUTIONS

Exercise 2.53

a) Here is a scatterplot of the data with year the explanatory variable and population the response.

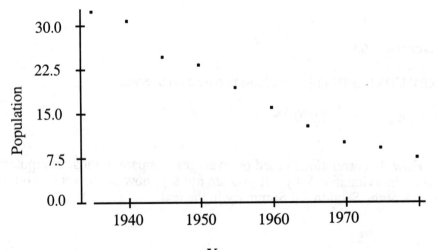

We find that the least-squares regression line is

Population = 1166.93 - 0.58679 x Year

b) The slope of the least-squares regression line indicates the decline in farm population per year over the period represented by the data. This is a decline of 0.58679 million people per year or 586,790 people per year. The percent of the observed variation in farm population accounted for by linear change over time is determined by the value of r^2, which is 0.977 (calculated using software). The desired percent is therefore 97.7%.

c) In 1990 the regression equation predicts the number of people living on farms to be about

$$\hat{y} = 1166.93 - 0.58679(1990) = -0.782 \text{ million.}$$

This result is unreasonable since population cannot be negative. This is an example of the dangers of extrapolation!

Exercise 2.55

Here are some additional possible lurking variables.

Amount of parental guidance or parental standards. Children whose parents do not set or enforce standards of behavior may engage in heavy TV watching and may spend little time on homework and studying (leading to poor grades).

Reading level. Children with poor reading skills do poorly in school. They may also spend more time watching TV rather than engaging in activities such as reading.

Behavioral or emotional problems. A child who relates poorly to others may tend to watch more TV (rather than interacting with others) and may do poorly in a classroom setting (has difficulty in settings that involve other people, such as a classroom).

There are undoubtedly other possibilities. Note that while these are possible lurking variables, additional study would need to be done to determine to what extent they explain any of the relationship between heavy TV watching and poor grades.

Exercise 2.63

a) We calculate $r^2 = (0.970)^2 = 0.9409$. This tells us that 94.09% of the observed variation in average SAT math scores is accounted for by linear change in average SAT verbal scores (or, equivalently, 94.09% of the observed variation in average SAT verbal scores is accounted for by linear change in average SAT math scores).

b) We would expect the correlation to be quite different, in fact lower. This is because correlations based on averages are usually too high when applied to individuals. Individuals exhibit more variation than averages; hence observed association is likely to be less strong with individuals.

Exercise 2.65

The explanatory variable is foreign language study. It has two values depending on whether a student has studied at least two years of a foreign language or has studied no foreign language. The response is the score on the English achievement test administered to seniors. Unfortunately for the study, students choose whether or not to take a foreign language. Academically stronger students are more likely than weaker students to study a foreign language. The academic strength of a student is the lurking variable here and prevents us from concluding that language study improves students' English scores.

SECTION 2.5

OVERVIEW

This section discusses techniques for describing the relationship between two or more categorical variables. To analyze categorical variables, we use counts (frequencies) or percents (relative frequencies) of individuals that fall into various categories. **A two-way table** of such counts is used to organize data about two categorical variables. Values of the **row variable** label the rows that run across the table, and values of the **column variable** label the columns that run down the table. In each cell (intersection of a row and column) of the table, we enter the number of cases for which the row and column variables have the values (categories) corresponding to that cell.

The **row totals** and **column totals** in a **two-way table** give the marginal distributions of the two variables separately. It is usually clearest to present these distributions as percents of the table total. **Marginal distributions** do not give any information about the relationship between the variables. **Bar graphs** are a useful way of presenting these marginal distributions.

The **conditional distributions** in a two-way table help us to see relationships between two categorical variables. To find the conditional distribution of the row variable for a specific value of the column variable, look only at that one column in the table. Express each entry in the column as a percent of the column total. There is a conditional distribution of the row variable for each column in the table. Comparing these conditional distributions is one way to describe the association between the row and column variables, particularly if the column variable is the explanatory variable. When the row variable is explanatory, find the conditional distribution of the column variable for each row and compare these distributions. Side-by-side bar graphs of the conditional distributions of the row or column variable can be used to compare these distributions and describe any association that may be present.

Data on three categorical variables can be presented as separate two-way tables for each value of the third variable. An association between two variables that holds for each level of this third variable can be changed, even reversed, when the data are combined by summing over all values of the third variable. **Simpson's paradox** refers to such reversals of an association.

GUIDED SOLUTIONS

Exercise 2.69

KEY CONCEPTS - marginal distribution

The marginal distribution of age can be found from the totals in each age group given in the bottom row of Table 2.9. Each value in this row must be divided by the total number of persons represented by the table found in the lower right corner. Now do the actual calculations, completing the following table.

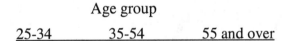

	Age group		
	25-34	35-54	55 and over
Fractions			
Percent			

Exercise 2.71

KEY CONCEPTS - describing relationships

Use the counts in the "Did not complete high school" row to find the requested percents. To organize your calculations, you may wish to fill in the following table.

	Age group		
	25-34	35-54	55 and over
Fractions			
Percent			

Now based on the percents you calculated, draw the bar graph.

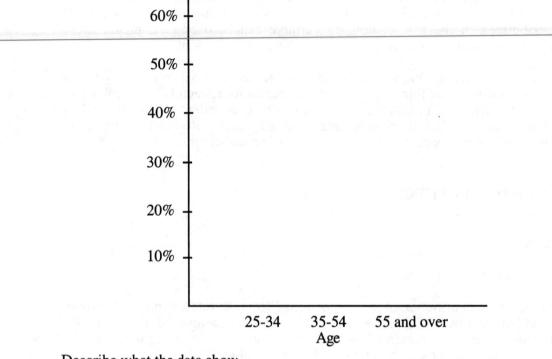

Describe what the data show.

Exercise 2.86

KEY CONCEPTS - marginal and conditional distributions, describing relationships

a) In considering the difference, recall that the table entries are in thousands of women. How were the table entries obtained?

b) To compute the marginal distribution, fill in the table.

	Never married	Married	Widowed	Divorced
Fraction				
Percent				

Now draw the bar graph.

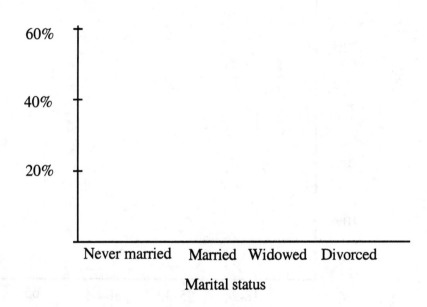

c) The conditional distributions are obtained by dividing the entry in each row by the total for the row and converting this fraction to a percent. Enter your results in the table.

	Never married	Married	Widowed	Divorced
18-24				
40-64				

Use the space for your description.

d) Fill in the percents corresponding to the conditional distribution of ages among single women.

	18-24	25-39	40-64	65
Percents				

Now display the distribution in a bar graph.

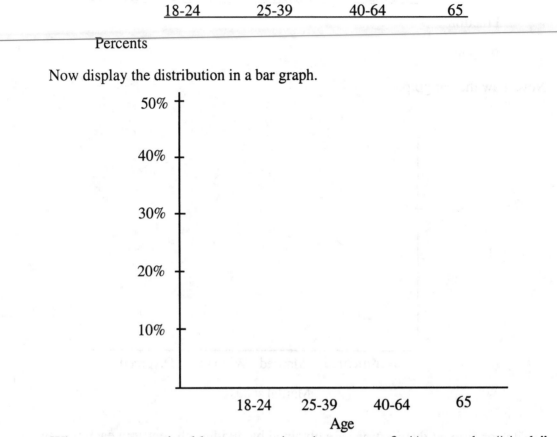

What age groups should your magazine aim to attract? (Assume that "single" means "never married.")

Exercise 2.87

KEY CONCEPTS - two-way tables, Simpson's paradox

a) Add corresponding entries in the two tables and enter the sums in the table.

	Admit	Deny
Male		
Female		

b) Convert your table in part a to one involving percentages of the row totals.

<u>Admit</u>   <u>Deny</u>

Male

Female

c) Now repeat the type of calculations you did in part b for each of the original tables.

Business

<u>Admit</u>   <u>Deny</u>

Male

Female

Law

<u>Admit</u>   <u>Deny</u>

Male

Female

d) To explain the apparent contradiction observed in part c, consider which professional school is easier to get into and which professional school males and females tend to apply to. Write your answer in plain English in the space provided. Avoid jargon and be clear!

COMPLETE SOLUTIONS

Exercise 2.69

The marginal distribution of age can be found from the totals in each age group given in the bottom row of Table 2.9. Each value in this row must be divided by the total number of persons represented by the table found in the lower right corner, which is 166,438. The result is

	Age group		
	25-34	35-54	55 and over
Fractions	$\dfrac{41,388}{166,438}$	$\dfrac{73,028}{166,438}$	$\dfrac{52,022}{166,438}$
Percent	24.86	43.88	31.26

Exercise 2.71

The percent of people in each age group who did not complete high school is determined by dividing the counts in the "Did not complete high school" row of the table by the total for this row (which is 30,512). The result is

	Age group		
	25-34	35-54	55 and over
Fractions	$\dfrac{5,325}{30,512}$	$\dfrac{9,152}{30,512}$	$\dfrac{16,035}{30,512}$
Percent	17.45	29.99	52.55

Following is a bar graph displaying these percents.

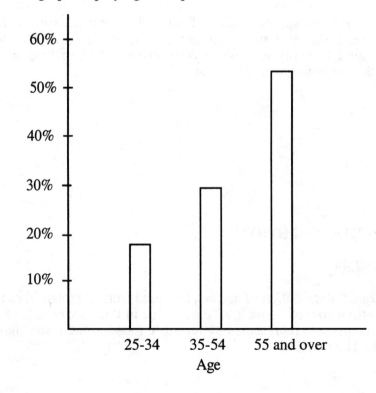

The data show that the older the age group, the higher the percentage who did not complete high school.

Although not requested, note that one possible explanation is that a greater emphasis on education (completion of high school) existed beginning in the 1950s (with concerns about the cold war and the space race) and this is reflected in the increasing percentage of people who did not complete high school with age.

Exercise 2.86

a) The sum of the entries in the Married column is actually

$$3,046 + 21,437 + 26,679 + 7,767 = 58,929$$

This differs from the entry in the Total row. This discrepancy is undoubtedly due to rounding off individual table entries to the nearest thousand women.

b) The distribution of marital status for all adult women is obtained by dividing the individual column totals by the table total given in the lower right corner of the table, then converting these fractions to percents.

	Never married	Married	Widowed	Divorced
Fraction	$\dfrac{19,312}{99,588}$	$\dfrac{58,931}{99,588}$	$\dfrac{11,080}{99,588}$	$\dfrac{10,266}{99,588}$
Percent	19.39	59.17	11.12	10.31

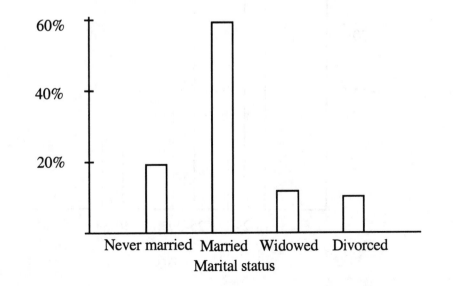

c) We give the conditional distributions, in percents, of marital status for both age groups. These percents were obtained by dividing the entry in each row by the total for the row and converting this fraction to a percent.

	Never married	Married	Widowed	Divorced
18-24	73.64%	24.15%	0.15%	2.06%
40-64	6.28%	72.67%	6.04%	15.00%

The major difference is one of experience with marriage. The majority of women 18-24 (73.64%) have never been married and very few are widowed or divorced. The great majority of women 40-64 (93.72%) are or have been married. These differences probably reflect the fact that marriage, widowhood, and divorce are more likely as one gets older.

d) The distribution of ages among single women (in percents) is obtained by dividing the entries in the "Never married" column by the total for that column (seen in the table to be 19,312). The result is

	18-24	25-39	40-64	65
Percents	48.10%	35.98%	11.95%	3.98%

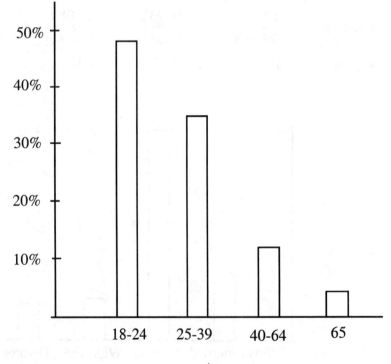

Obviously, the percent of single women decreases as age increases ,and it makes sense to target women under 40. Of the women who have never been married, 84.08% are below age 40.

Exercise 2.87

a) Here is the desired two-way table.

	Admit	Deny
Male	490	210
Female	280	220

b) We first add a column containing the row totals to the table in part a.

	Admit	Deny	Total
Male	490	210	700
Female	280	220	500

We now convert the table entries to percents of the row totals. We divide the entries in the first row by 700 and express the results as percents. We divide the entries in the second row by 500 and express these as percents.

	Admit	Deny
Male	70%	30%
Female	56%	44%

We see that Wabash admits a higher percent of male applicants.

c) We repeat the calculations in part b, but for each of the original two tables.

Business

	Admit	Deny	Total
Male	480	120	600
Female	180	20	200

Law

	Admit	Deny	Total
Male	10	90	100
Female	100	200	300

Converting entries to percents of the row totals yields

	Business	
	Admit	Deny
Male	80%	20%
Female	90%	10%

	Admit	Deny
Male	10%	90%
Female	33.3%	66.7%

We see that each school admits a higher percentage of female applicants.

d) Although both schools admit a higher percentage of female applicants, the admission rates are quite different. Business admits a high percentage of all applicants; it is easier to get into the business school. Law admits a lower percentage of applicants; it is harder to get into the law school Most of the male applicants to Wabash apply to the business school with its easy admission standards. Thus, overall, a high percentage of males are admitted to Wabash. The majority of female applicants apply to the law school. Because it has tougher admission standards, this makes the overall admission rate of females appear low, even though more females are admitted to both schools!

SELECTED TEXT REVIEW EXERCISES

GUIDED SOLUTIONS

Exercise 2.91

KEY CONCEPTS - Drawing and interpreting a scatterplot with a categorical variable, unusual observations in a scatterplot

a) The explanatory variable would be age in this example. Why? Use the axes given to make your scatterplot. For simplicity, consider using a circle for survivors and a plus for those who died.

Note: Most software that will draw a scatterplot will allow you to use different symbols for the points based on a third variable - all three variables are entered as part of the data set. If you are using a software package for this course, try to use it to make a plot with axes similar to the one here.

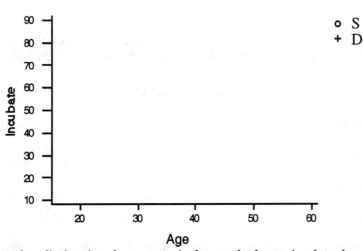

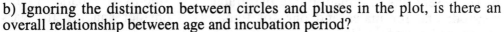

b) Ignoring the distinction between circles and pluses in the plot, is there an overall relationship between age and incubation period?

c) There are several things to look for here. Is there a different relationship between age and incubation for those who died and those who survived? Are the survivors different from those who died in terms of incubation period or age - that is, did the survivors tend to be younger or have longer incubation periods?

d) Unusual observations can be in the *x* variable or the *y* variable or they can be any observations that don't follow the general pattern of the data. There isn't much of a pattern in the data, so just look for unusual observations in terms of age or incubation. In which variable would it be more interesting to find an unusual observation?

Exercise 2.101

KEY CONCEPTS - conditional distributions in two-way tables, side-by-side bar graphs

Begin by converting the counts in the table to percentages of the column totals (in other words, compute the conditional distributions given gender).

	Male	Female
Firearms		
Poison		
Hanging		
Other		

Next, present this information in a bar graph. Use side-by-side bars for each method; one for males and one for females. Consider shading the bars corresponding to females.

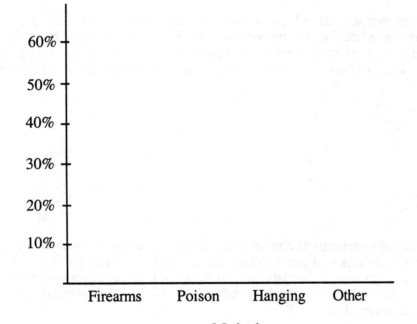

Now write a brief account of differences in suicide between men and women.

COMPLETE SOLUTIONS

Exercise 2.91

a) Graph given.

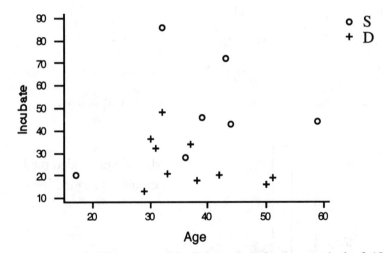

Note: There are two deaths at age 32 with an incubation period of 48 hours, so there are only 17 distinct points on the graph instead of 18. Find the plus in the graph corresponding to these two equal observations.

b) There doesn't appear to be an overall relationship between age and incubation.

c) If you examine just the circles or just the pluses, there doesn't appear to be a relationship between age and incubation for those who died or those who survived. The thing that stands out in the scatterplot is that survival appears to be related to the incubation period. If the incubation period was longer the person seems more likely to have survived regardless of age. For the incubation periods of 21 hours or less, 6 out of 7 people died, while for incubation periods exceeding 21 hours, there are only 5 deaths out of 11 people (remember the double point if counting from the graph).

d) There isn't much of a pattern in the data, so just look for unusual observations in terms of age or incubation. There is one person who is younger than most (17 years) but that doesn't suggest much. There are two people with incubation periods that are much longer than the rest (72 and 86 hours) and we might want to examine those points a little further. Is there something that occurred differently to those two individuals, possibly with respect to their exposure?

Exercise 2.101

We convert the table to percents of males and females by dividing by the column totals.

	Male	Female
Firearms	64.45%	41.99%
Poison	14.04%	34.62%
Hanging	15.05%	13.17%
Other	6.46%	10.22%

A bar graph that presents the same information is given below.

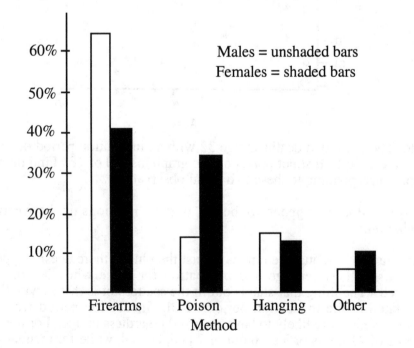

A few things are striking. About four times as many men (25,415) as women (6,095) commit suicide. Our prior belief was that women were more likely to commit suicide than men. Firearms are the most popular method for both sexes, but men use firearms more frequently than women. Poison is nearly as popular as firearms for women. Poison is not as popular for men as it is for women. This may suggest that women have a greater preference for slower, more passive methods (such as poison) than men. Men seem to prefer quicker methods (firearms).

CHAPTER 3

PRODUCING DATA

SECTION 3.1

OVERVIEW

Data can be produced in a variety of ways. **Sampling**, when done properly, can yield reliable information about a **population**. **Observational studies** are investigations in which one simply observes the state of some population, usually with data collected by sampling. Even with proper sampling, data from observational studies are generally not appropriate for investigating cause-and-effect relations between variables. **Experiments** are investigations in which data are generated by active imposition of some treatment on the subjects of the experiment. Properly designed experiments are the best way to investigate cause-and-effect relations between variables.

The **population** is the entire group of individuals or objects about which we want information. The information collected is contained in a **sample**, the part of the population we actually observe. How the sample is chosen (the sampling **design**) has a large impact on the usefulness of the data. A useful sample is representative of the **population** and helps us answer our questions. "Good" methods of collecting a sample include the following:

> **Simple random samples (SRS)**
> **Probability samples**
> **Stratified random samples**
> **Multistage samples**

All these sampling methods involve some aspect of randomness through the use of a formal chance mechanism. Random selection is just one precaution a person can take to reduce **bias,** the systematic favoring of a certain outcome. Samples we select using our own judgment, because they are convenient, or "without forethought" (mistaking this for randomness) are usually biased. This is why we use computers or a **table of random digits** to help us select a sample.

There are many ways to choose a **simple random sample (SRS)**. An SRS of size n is a collection of n individuals chosen from the population in a manner so that each possible set of n individuals has an equal chance of being selected. In practice, simple methods such as drawing names from a hat is one way of getting an SRS. The names in the hat are the units in the population. To choose the sample, we mix up the names and select the sample of n at random from the hat. In reality the population may be very large. A computer or a table of random numbers can be used to mimic the process of pulling names from a hat.

The method of selecting an SRS using a table of random digits can be summed up in two steps.

   1. Give every individual in the population its own numerical label. All labels need to have the same number of digits.
   2. Starting anywhere in the table (usually a spot selected at random), read off labels until you have selected as many labels as needed for the sample.

Another common type of sample design is a **stratified random sample**. Here the population is first divided into **strata** and then an SRS is chosen from each stratum. The strata are formed using some known characteristic of each individual thought to be associated with the response to be measured. Examples of strata are gender or age. Individuals in a particular stratum should be more like one another than those in the other strata.

Poor sample designs include the **voluntary response sample**, where people place themselves in the sample, and the **convenience sample**. Both of these methods rely on personal decision for the selection of the sample; this is generally a guarantee of bias in the selection of the sample.

Be on the lookout for other kinds of bias that can occur even in well designed studies:

   Nonresponse bias which occurs when individuals who are selected do not participate or cannot be contacted
   Bias in the **wording of questions** leading the answers in a certain direction
   Confounding or confusing the effect of two or more variables
   Undercoverage, which occurs when some group in the population is given either no chance or a much smaller chance than other groups to be in the sample
   Response bias which occurs when individuals do participate but are not responding truthfully or accurately due to the way the question is worded,

the presence of an observer, fear of a negative reaction from the interviewer, or any other such source.

These types of bias can occur even in a randomly chosen sample, and we need to try to reduce their impact as much as possible.

GUIDED SOLUTIONS

Exercise 3.7

KEY CONCEPTS - selecting an SRS with a table of random numbers

The table of random numbers can be used to select an SRS of numbers. In order to use it to select a random sample of six minority managers, the managers need to be assigned numerical labels. So that everyone does the problem the "same" way, we first label the managers according to alphabetical order.

01 - Agarwal	08 - Dewald	15 - Huang	22 - Puri
02 - Anderson	09 - Fernandez	16 - Kim	23 - Richards
03 - Baxter	10 - Fleming	17 - Liao	24 - Rodriguez
04 - Bowman	11 - Gates	18 - Mourning	25 - Santiago
05 - Brown	12 - Goel	19 - Naber	26 - Shen
06 - Castillo	13 - Gomez	20 - Peters	27 - Vega
07 - Cross	14 - Hernandez	21 - Pliego	28 - Wang

If you go to line 139 in the table and start selecting two-digit numbers, then you should get the same answer as given in the complete solution. If your entire sample has not been selected by the end of line 139, continue to the next line in the table.

The sample consists of the managers:

Exercise 3.11

KEY CONCEPTS - stratified random sample

What are the two strata from which you are going to sample? A stratified random sample consists of taking an SRS from each stratum and combining these to form the full sample. How large an SRS will be taken from each of the two strata? How would you label the units in each of the two strata from which you will sample? Fill in the following table to describe your sampling plan.

	Strata	
	Midsize accounts	Small accounts
Number of units in stratum		
Sample size		
Labeling method		

In practice, you would probably use the table of random numbers to first select the SRS from the midsize accounts and then you would select the SRS from the small accounts. In this problem you are going to select not the full samples but only the first five units from each stratum. Start in Table B at line 115 and first select five midsize accounts and then continue in the table to select five small accounts. Write down the numerical labels.

First five midsize accounts:

First five small accounts:

Exercise 3.13

KEY CONCEPTS - sampling frame, undercoverage

a) Which households wouldn't be in the sampling frame? Make some educated guesses as to how these households might differ from those in the sampling frame (other than the fact that they don't have a phone number in the directory).

b) Random-digit dialing makes the sampling frame larger. Which households are added to it?

Exercise 3.15

KEY CONCEPTS - wording of questions

Questions can be worded to make it seem as though any reasonable person should agree (disagree) with the statement. Even though both (A) and (B) address the same issue, the tone of each question is different enough from the other to elicit a different response. Try to decide which question should have a higher proportion favoring banning contributions.

Exercise 3.17

KEY CONCEPTS - explanatory and response variables, experiments and observational studies

What are the researchers trying to demonstrate with this study? What groups are being compared and did the experiment deliberately impose membership in the groups on the subjects to observe their responses? What are the explanatory and response variables?

Explanatory variable:

Response variable:

Observational study or experiment? (Circle one.)

Exercise 3.21

KEY CONCEPTS - voluntary response sampling, bias

This is an example of a call-in opinion poll in which the subjects select themselves to be in the sample. In this case, which subjects do you think are more likely to be included in the sample and in which direction would this bias the results?

Exercise 3.23

KEY CONCEPTS - populations, samples, bias

What is the population of interest to the Miami Police Department? Is there a problem with undercoverage in the way the addresses for the sample are selected? How might this bias the results? Could there be response bias in this study as well? How might this bias the results?

Exercise 3.25

KEY CONCEPTS - selecting an SRS with a table of random numbers

The table of random numbers can be used to select an SRS of blocks. In order to use it to sample blocks from the census tract, the blocks need to be assigned numerical labels. Since the blocks are already assigned three-digit numbers (the numbers do not need to be consecutive to select a SRS), enter Table B at line 125 and select three-digit numbers ignoring those that don't correspond to blocks on the map. If your entire sample has not been selected by the end of line 139, continue to the next line in the table.

The sample consists of the blocks numbered:

Exercise 3.31

KEY CONCEPTS - inference about the population, sample size

The margin of error for all adults is three percentage points in either direction, while for men the margin of error is five percentage points in either direction. What is the effect of sample size on the accuracy of the results? How does this explain the difference in the two margins of error?

COMPLETE SOLUTIONS

Exercise 3.7

To choose an SRS of six managers to be interviewed, first label the members of the population by associating a two-digit number with each.

01 - Agarwal	08 - Dewald	15 - Huang	22 - Puri
02 - Anderson	09 - Fernandez	16 - Kim	23 - Richards
03 - Baxter	10 - Fleming	17 - Liao	24 - Rodriguez
04 - Bowman	11 - Gates	18 - Mourning	25 - Santiago
05 - Brown	12 - Goel	19 - Naber	26 - Shen
06 - Castillo	13 - Gomez	20 - Peters	27 - Vega
07 - Cross	14 - Hernandez	21 - Pliego	28 - Wang

Now enter Table B at line 139 and read two-digit groups until six managers are chosen.

55|58|8   9|94|04| 70|70|8   4|<u>10</u>|98   43|56|3   5|69|34| 48|39|4   5|<u>17</u>|<u>19</u>|
<u>12</u>|97|5   1|32|58| <u>13</u>|04|8

The selected sample is 04 - Bowman, 10 - Fleming, 17 - Liao, 19 - Naber, 12 - Goel, and 13 - Gomez.

Exercise 3.11

There are 500 midsize accounts. We are going to sample 5% of them, which is 25. Label the accounts 001, 002, ..., 500 and select an SRS of 25 of the midsize accounts. There are 4400 small accounts. We are going to sample 1% of them, which is 44. Label the accounts 0001, 0002, ..., 4400 and select an SRS of 44 of the small accounts.

Starting at line 115, we first select five midsize accounts, that is, an SRS of size five, using the labels 001 through 500. Continuing in the table we select five small accounts, that is, an SRS of size five, using the labels 0001 through 4400. Note that for the midsize accounts we read from Table B using three-digit numbers, and for the small accounts we read from the table using four-digit numbers.

610|<u>41</u>   7|768|<u>4</u>   94|<u>322</u>| <u>247</u>|09   7|<u>3698</u>| <u>1452</u>|6   318|9332|592
1|<u>4459</u>| <u>2605</u>|6   314|<u>24</u>   80|<u>371</u>   6|

The first five midsize accounts are those with labels 417, 494, 322, 247, and 097. Continuing in the table, using four digits instead of three, the first five small accounts are those with labels 3698, 1452, 2605, 2480, and 3716.

Exercise 3.13

a) Households omitted from the frame are those that do not have a telephone number listed in the telephone directory. The types of people who might be underrepresented are poorer (including homeless) people who cannot afford to have a phone and people who have unlisted numbers. It is harder to characterize the second group. As a group they would tend to have more money, because you need to pay to have your phone number unlisted; it might also include more single women who do not want their phone numbers available and possibly people whose jobs put them in contact with large groups of people who might harass them if their phone number were easily accessible.

b) People with unlisted numbers will be included in the sampling frame. The sampling frame would now include any household with a phone. One interesting point is that all households will not have the same probability of getting into the sample, as some households have more than one phone line and will be more likely to get into the sample. So, strictly speaking, random-digit dialing will provide not an SRS of households with phones but an SRS of phone numbers!

Exercise 3.15

As with many pairs of questions, changes in wording of the "same" question can elicit different responses.

(A)   The wording implies that contributions from special interest groups are bad - huge sums of money will exchange hands for the benefit of the special interest groups.

(B)   The wording suggests that we would be undemocratic to deny special interest groups the same right to contribute money as private individuals and other groups.

We would guess that 80% favored banning contributions when presented with question (A) and 40% favored banning contributions when presented with question (B).

Exercise 3.17

The researchers are hoping to study the effect of living in public housing on family stability in poverty-level households. The two groups being studied are those that were accepted into public housing and those that weren't during the previous year. The researchers did not impose the treatment (living in public housing) on the subjects, so this is an observational study. The explanatory variable is whether or not they lived in (were accepted into) public housing, and the response is the measure of family stability being used.

Since family stability or variables related to it were probably used in the admission process for public housing, families accepted for public housing probably were more stable before they entered public housing than were those turned down. Because of these possible lurking variables, we could not attribute differences in family stability to living in public housing.

Exercise 3.21

A voluntary response sample is almost guaranteed to yield a sample that doesn't represent the entire population. A voluntary sample typically displays bias, or systematic error, in favoring some parts of the population over others. Often it is possible to use common sense to figure out the likely direction of bias. Are subjects who favor tougher gun laws more likely to call? This is possible, because they may be the ones dissatisfied with current laws and more likely to want to make their opinions known.

Note that the direction of bias is not as clear-cut here as in some other examples. It might have been your educated guess that those who like the current legislation would be more likely to call in as they don't want the law changed, which would cause bias in the other direction. There is nothing wrong with having this opinion. The main thing is to realize that bias is likely to be present in this example and to recognize why it would occur.

Exercise 3.23

The population of interest is black residents of Miami. The sample is the black residents at the 300 mailing addresses chosen by the police. (If a nonblack resident lives at one of these addresses, their opinions would not be included in the sample, as they are not part of the population).

The police are interviewing residents at addresses in predominantly black neighborhoods. The opinions of black residents who do not live in predominantly black neighborhoods could not be included in the sample, resulting in undercoverage. Black residents in other neighborhoods are likely to feel differently about the police, creating a potential source of bias if the population of interest is really *all* black residents. These black residents of Miami may be more satisfied with police service.

A more serious source of bias is response bias. It is unlikely adults will be honest when giving opinions of police service to a police officer. They will probably indicate a greater degree of satisfaction than they really feel.

Exercise 3.25

The random number table starting at line 125 is reproduced below. Using three-digit numbers, select only those corresponding to blocks in the tract. The five underlined blocks, 214, 313, 409, 306, and 511, correspond to the sample.

```
967I46  1I214I9  37I823I  718I68  1I844I2  35I119I  621I03  3I024I4
96I927I  199I31  3I680I9  74I192I  775I67  8I874I1  48I409I  419I03
4I390I9  99I477I  253I30  6I435I9  40I085I  169I25  8I511I7  36I071
```

Exercise 3.31

Larger random samples give more accurate results than smaller sample sizes, so results based on a larger sample have smaller margins of error. Since the men are only about one-third of the sample of all adults, we will know less about the subgroup of men than we do about the group of all adults (a sample size of 472 versus a sample size of 1025 + 472 = 1397).

SECTION 3.2

OVERVIEW

Experiments are studies in which one or more **treatments** are imposed on experimental **units** or **subjects**. A treatment is a combination of **levels** of the explanatory variables, called **factors**. The design of an experiment is a specification of the treatments to be used and the manner in which units or subjects are assigned to the treatments. The basic features of well-designed experiments are **control**, **randomization**, and **replication**.

Control is used to avoid confounding (mixing up) the effects of treatments with the effects of other influences, such as lurking variables. One such lurking variable is the **placebo effect**, which is the response of a subject to the fact of receiving any treatment. The simplest form of control is **randomized comparative experimentation**, which involves comparisons between two or more treatments. One of these treatments may be a **placebo** (fake treatment); subjects receiving the placebo are referred to as a **control group**.

Randomization can be carried out using the ideas we learned in Section 3.1. Randomization is carried out before applying the treatments and helps control bias by creating treatment groups that are similar. Replication, the use of many units in an experiment, is important because it reduces the chance variation between treatment groups arising from randomization. Using more units helps increase the ability of your experiment to establish differences between treatments.

Further control in an experiment can be achieved by forming experimental units into **blocks** that are similar in a way that is thought to affect the response, similar to strata in a stratified sample design. In a **block design**, units are first formed into blocks and then randomization is carried out separately in each block. **Matched pairs** are a simple form of blocking used to compare two treatments. In a matched pairs experiment either the same unit (the block) receives both treatments in a random order or very similar units are matched in pairs (the blocks). In the latter case, one member of the pair receives one of the treatments and the other member the remaining treatment. Members of a matched pair are assigned to treatments using randomization.

Some additional problems that can occur that are unique to experimental designs are **lack of blinding** and **lack of realism**. These problems should be addressed when designing the experiment.

GUIDED SOLUTIONS

Exercise 3.35

KEY CONCEPTS - lurking variables, randomized comparative experiments

a) Are there any variables that could be different between the surgical and nonsurgical treatment groups and also be related to the increased death rate? These would be lurking variables.

b) Using the examples in the text, diagram a completely randomized design for the study, using labels that are appropriate for this study. Indicate the size of the treatment groups and a possible response variable.

Exercise 3.41

KEY CONCEPTS - double-blind experiments, bias

What does it mean for the ratings to be blind? Was this done in this case? If not, how can this bias the results and in which direction?

Exercise 3.43

KEY CONCEPTS - matched pairs design, randomization

The first thing you should do is identify the treatments and the response variable. Next decide what are the matched pairs in this experiment. How will you use a coin flip to assign members of a pair to the treatments? What will you measure and how will you decide whether the right hand tends to be stronger in right-handed people?

Exercise 3.47

KEY CONCEPTS - identifying experimental units or subjects, factors, treatments, and response variables, completely randomized design, randomization

a) Read the description of the study carefully. To identify the subjects, ask yourself exactly who was used in the study.

b) To identify the factor and its levels, ask yourself what question the experiment wished to answer? What did they vary to answer the question, and at how many levels?

    Factor:

    Levels:

    Response variable:

c) Using the examples in the text, diagram a completely randomized design for the study, using labels that are appropriate for this study.

d) Label the children 001 to 210. Enter Table B at line 125 and select an SRS of 70 to receive Set 1. Continue in Table B, selecting 70 more to receive Set 2. The remaining 70 receive Set 3. Show that you can carry this out by identifying the first five children assigned to the first treatment (Set 1).

    First five children assigned to Set 1:

Exercise 3.53

KEY CONCEPTS - design of an experiment, randomization

a) Since the hypothesis is that calcium will reduce blood pressure, you should measure all subjects' blood pressure before starting the experiment. You must also decide on a suitable amount of time to run the experiment. It should be sufficiently long to allow the treatment to show some effect. After the allotted time has elapsed, the blood pressure of all subjects should be measured again. How do you know a reduction in blood pressure is not due to the placebo effect? Any potential problem caused by the placebo effect should be taken care of in the design. Design an appropriate experiment on these 40 men that takes the placebo effect into account. You can outline your design in words or a diagram.

b) Label the 40 names using two-digit labels. To have your answer agree with the complete solution, start with the label 01 and label down the columns. (Of course, one could start with another number or label across rows if one wished.) We need to select an SRS of 20 subjects to receive the high-calcium diet with the remainder to receive the placebo. The names are

Alomar	Denman	Han	Liang	Rosen
Asihiro	Durr	Howard	Maldonado	Solomon
Bennett	Edwards	Hruska	Marsden	Tompkins
Bikalis	Farouk	Imrani	Moore	Townsen
Chen	Fratianna	James	O'Brian	Tullock
Clemente	George	Kaplan	Ogle	Underwood
Cranston	Green	Krushchev	Plochman	Willis
Curtis	Guillen	Lawless	Rodriguez	Zhang

Now start reading line 121 in Table A. You will need to keep reading until you have selected all the names for those in the group to receive calcium supplements. This may require you to continue on to line 122 and subsequent lines. Since there are only two groups in this experiment, you can stop after you have selected the names for those to receive the calcium supplements. The remaining names are assigned to receive the placebo.

Subjects assigned to receive calcium supplements:

Exercise 3.55

KEY CONCEPTS - placebo effect

What is the placebo effect? Does it imply that there was no physical basis for the patients' pain, or does it refer to something else?

Exercise 3.59

KEY CONCEPTS - block design, sample size

a) This is an example of a block design. Remember, a block is a group of experimental units known before the experiment to be similar in some way that is expected to affect the response to the treatments. What are the blocks in this experiment? The random assignment of units to treatments is carried out separately within each block. Your graphical representation of the experiment should illustrate the blocks and the randomization.

b) Reread the end of Section 3.1, which discusses larger sample sizes. How do these ideas apply here?

COMPLETE SOLUTIONS

Exercise 3.35

a) The more seriously ill patients may be assigned to the new method by their doctors. This might be done because experience tells the doctors that there is little chance of surgery being successful in the more serious cases and little harm in trying the new treatment on them. The seriousness of the illness would be a lurking variable that would make the new treatment look worse than it really is.

b)

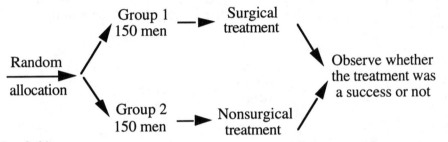

Exercise 3.41

The ratings were not blind because the experimenter who rated their level of anxiety presumably knew whether they were in the meditation group or not. Since the experimenter was hoping to show that those who meditated had lower levels of anxiety, he might unintentionally rate those in the meditation group as having lower levels of anxiety if there is any subjectivity in the rating. It would be better if a third party who did not know which group the subjects belonged to rated the subjects' anxiety levels.

Exercise 3.43

Ten subjects are available. There are two treatments in the study. Treatment 1 is squeezing with the right hand and treatment 2 is squeezing with the left hand. The response is the force exerted as indicated by the reading on the scale.

To do the experiment, we use a matched pairs design. The matched pairs are the two hands of a particular subject. We should randomly decide which hand to use first, perhaps by flipping a coin. We measure the response for each hand and then compare the forces for the left and right hands over all subjects to see if there is a systematic difference between the two hands.

Exercise 3.47

a) The subjects are the 210 children aged 4 to 12 years who participated in the study. The problem tells us nothing about how these children were selected to be in the study.

b) The factor is the set of choices that are presented to each subject. The levels correspond to the three sets of choices, Set 1, Set 2 and Set 3, so there are three levels in this problem. (You might have thought the levels were the number of choices presented to the subject, but sets 2 and 3 both have six choices and correspond to different levels of the factor). The response variable is whether they chose a milk drink or a fruit drink. Since they had to choose one or the other, the response could just be whether or not they chose a milk drink (yes or no).

c)

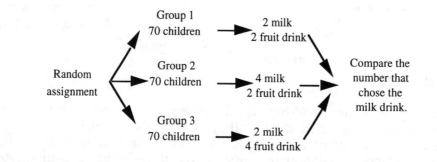

d) Table B starting at line 125 is reproduced here. Read across the row in groups of digits equal to the number of digits you used for your labels (for example, since you used three digits for labels, read line 125 in triples of digits). The five underlined numbers 119, 033, 199, 192, and 148 represent the first five children to be assigned to receive Set 1.

96746 12149 37823 71868 18442 35119 62103 39244
96927 19931 36089 74192 77567 88741 48409 41903

Exercise 3.53

a) Measure all subjects' blood pressure before starting the experiment. Assign 20 subjects at random to the added calcium diet and the remainder to the placebo group. After a suitable amount of time has passed for the treeatment to show some effect, measure the blood pressure again and compare the change for the two groups. If the calcium is effective, the calcium group should show a *greater* lowering of blood pressure than the placebo group. (Remember that the placebo group might also show some lowering of their blood pressure. That's the placebo effect.) In a diagram, the design is represented as

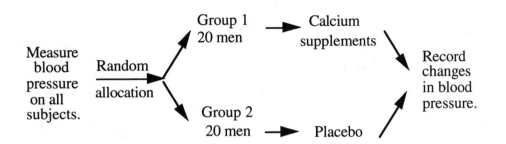

b) The names with their labels are

01 Alomar	09 Denman	17 Han	25 Liang	33 Rosen
02 Asihiro	10 Durr	18 Howard	26 Maldonado	34 Solomon
03 Bennett	11 Edwards	19 Hruska	27 Marsden	35 Tompkins
04 Bikalis	12 Farouk	20 Imrani	28 Moore	36 Townsen
05 Chen	13 Fratianna	21 James	29 O'Brian	37 Tullock
06 Clemente	14 George	22 Kaplan	30 Ogle	38 Underwood
07 Cranston	15 Green	23 Krushchev	31 Plochman	39 Willis
08 Curtis	16 Guillen	24 Lawless	32 Rodriguez	40 Zhang

Now start reading line 119 in Table B. Read across the row in groups of digits equal to the number of digits you used for your labels (since you used two digits for labels, read line 119 in pairs of digits).

95857   07118   87664   92099   58806   66979   98624   84826
35476   55972   39421   65850   04266   35435   43742   11937
71487   09984   29077   14863   61683   47052   62224   51025
13873   81598   95052   90908   73592   75186   87136   95761

Those assigned to the calcium supplemented diet are 18 - Howard, 20 - Imrani, 26 - Maldonado, 35 - Tompkins, 39 - Willis, 16 - Guillen, 04 - Bikalis, (26 and 35 are the next two numbers but both are skipped because they have been selected already), 21 - James, 19 - Hruska, 37 - Tullock, 29 - O'Brian, 07 - Cranston, 34 - Solomon, 22 - Kaplan, 10 - Durr, 25 - Liang, 13 - Fratianna, 38 - Underwood, 15 - Green, and 05 - Chen. The remaining men are in the placebo group.

Exercise 3.55

Many patients respond favorably to any treatment, even a placebo that has no therapeutic effect. There could have been a physical basis for the patients' pain and they are responding to the fact of having received *any* treatment, which is the meaning of the placebo effect.

Exercise 3.59

a)

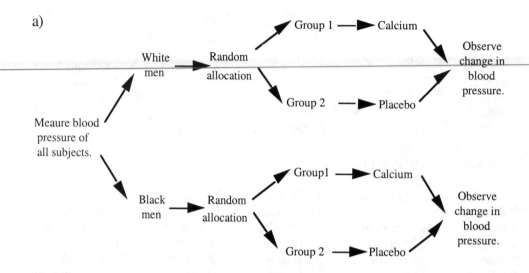

b) A larger group of subjects will provide more information because individual differences tend to be more evenly balanced between the treatments (all those with high blood pressure are less likely to be in the same group). When comparing sample averages between the treatments, the averages will be more likely to reflect treatment differences and be less influenced by individual differences.

SELECTED TEXT REVIEW EXERCISES

GUIDED SOLUTIONS

Exercise 3.61

KEY CONCEPTS - observational studies and experiments

What are the groups or "treatments" in this study? How were the subjects assigned to the groups? Was a treatment deliberately imposed on individuals to observe their response? It may help to think about what the explanatory and response variables are in this example.

Exercise 3.63

KEY CONCEPTS - populations and samples, sample size, bias

a) Try to identify the population as exactly as possible. Where the information is incomplete you may need to make assumptions to try and describe the population in a reasonable way. What is the sample in this example?

b) Are the sample sizes large enough to represent the population fairly accurately? Is there a possibility of response bias in this example?

Exercise 3.69

KEY CONCEPTS - explanatory and response variables, factors and levels, completely randomized design

a) The individuals on which an experiment is done are the experimental units. What are the experimental units in this study? What response is being observed?

Experimental units:

Response:

b) The explanatory variables in an experiment are the factors. How many factors are there in this experiment? When there is more than one factor in an experiment, a "treatment" is formed by combining a level of each of the factors. How many levels of each factor are there, and how many treatments can be formed? If there are 10 chicks for each treatment, how many experimental units are needed? Use a diagram to illustrate the different combinations of the levels of the factors to form the treatments.

c) The experiment is run as a completely randomized design with how many treatments? Diagram the design for this experiment.

COMPLETE SOLUTIONS

Exercise 3.61

The subjects are all executives who have volunteered for an exercise program. They take a physical examination that is used to divide them into two groups corresponding to low fitness and high fitness. The experimenter did not impose the level of physical fitness on the subjects. Their level of physical fitness was already present and it, not the experimenter, placed them in the groups. The experimenter had no role in who was assigned to low fitness and high fitness. Those who are in the low fitness group may be heavier, so weight could be a lurking variable that might explain differences in leadership.

This is an observational study. The individuals were observed and the variables of interest - physical fitness and leadership - were measured. There was no attempt to influence the response variable leadership by altering the subjects' level of physical fitness.

Exercise 3.63

a) The population is Ontario residents. From the problem, there is no reason to believe that the population is restricted to adult residents or any other subgroup. Any person residing in Ontario is in the population. The sample is the 61,239 people who were interviewed.

b) The sample size is quite large, and in an SRS you would suspect about half the sample to be men and half to be women, so there should be a reasonably large sample size from each of the groups. This should give you estimates that are pretty close to the truth about the entire population unless there is bias present.

This example is similar to the example used in the discussion of response bias. Answers to questions that ask respondents to recall past events are often inaccurate because of faulty memory. People may bring past events forward to more recent time periods, so someone who had visited a general practitioner 14 months ago might answer yes to this question. These estimates might be larger than the true percentage due to response bias. Good interviewing techniques can reduce this source of bias.

Exercise 3.69

a) The one-day old male chicks are the experimental units. They are the objects that receive the treatment. The response variable is the amount of weight they gain in 21 days, or the difference between their initial weight and their weight after three weeks.

b) This is an example of an experiment with two factors. The first factor is the variety of corn, which has three levels: normal, opaque-2 and floury-2. The second factor is the protein level of the diet, which has three levels: 12% protein, 16% protein, and 20% protein. A diet or treatment is made up of a combination of a type of corn and a level of protein, so there are 3 x 3 = 9 possible treatments or diets. Since 10 chicks are assigned to each diet, the experiment needs 9 x 10 = 90 experimental units. The table shows the layout of the treatments.

		Factor B Variety		
		Normal	Opaque-2	Floury-2
	12%	1	2	3
Factor A % protein	16%	4	5	6
	20%	7	8	9

Combinations of the levels of the two factors form nine treatments.

c) In the following diagram, group 1 gets treatment 1, group 2 gets treatment 2, and so on, where the treatment numbers correspond to those in part b. There are nine treatments and nine groups.

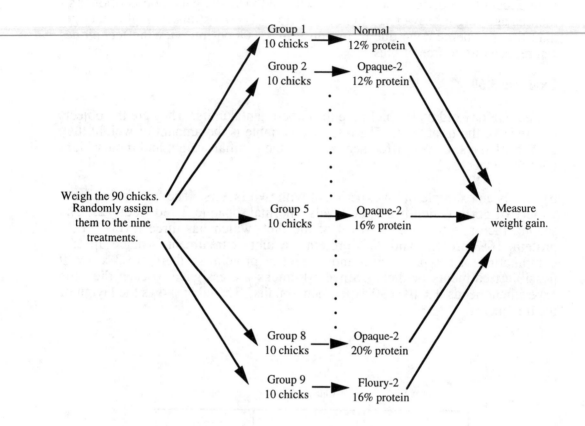

CHAPTER 4

PROBABILITY AND SAMPLING DISTRIBUTIONS

SECTION 4.1

OVERVIEW

This section introduces **statistical inference**, **parameters**, and **statistics.** [*population* handwritten above "parameters"; *sample* handwritten above "statistics"] Statistical inference is the technique that allows us to use the information in a sample to draw conclusions about the population. To understand the idea of statistical inference, it is important to understand the distinction between parameters and statistics. A **statistic** is a number we calculate based on a sample from the population - its value can be computed once we have taken the sample, but its value varies from sample to sample. This variation from sample to sample is called **sampling variability**. A statistic is generally used to estimate a population **parameter**, a fixed but unknown number that describes the population. In the real world we rarely know the population parameters and we need statistics to estimate them.

We will make estimates about parameters using the rules of **probability**. Probability is a branch of mathematics that studies random processes over a long series of repetitions. A process or phenomenon is called **random** if its outcome is uncertain. Although individual outcomes are uncertain, when the process is repeated a large number of times the underlying distribution for the possible outcomes begins to emerge. For any outcome, its **probability** is the proportion of times, or the relative frequency, with which the outcome would occur in a long series of repetitions of the process. It is important that these repetitions or trials be **independent** for this property to hold. By independence we mean that the outcome of one trial must not influence the outcome of any other.

You can study random behavior by carrying out physical experiments such as coin tossing or die rolling, or you can simulate a random phenomenon on the computer. Using the computer is particularly helpful when we want to consider a large number of trials. Randomness and independence are the keys to using the rules of probability.

GUIDED SOLUTIONS

Exercise 4.1

KEY CONCEPTS - statistics and parameters

In deciding whether a number represents a parameter or a statistic, you need to think about whether it is a number that describes a population of interest or whether it is a number computed from the particular sample that was selected. What is the population and what is the sample in the problem? Based on this, indicate whether the number is a parameter or a statistic.

> 2.5003 cm: parameter
>
> 2.5009 cm: statistic

Exercise 4.7

KEY CONCEPTS - estimating probabilities as the proportion of times an event occurs in many repeated trials

a) Record the number of tosses until the first head for your 50 trials.

The probability of an event is the proportion of times the event occurs in many repeated trials of a random phenomenon. To estimate the probability of a head on the first toss, determine the proportion of times this occurred in your 50 trials.

Estimate =

What value should you expect this probability to be for a fair coin?

b) Now use the results of your 50 tosses to determine the proportion of times the first head appeared on an odd-numbered toss.

Estimate =

Exercise 4.9

KEY CONCEPTS - interpreting probability

The probability of an event is the proportion of times the event occurs in many repeated trials of a random phenomenon. What is the event of interest in this problem? What is the random phenomenon in this problem? What is a trial? Now state the meaning of the probability 1/50 in this situation in simple language.

Exercise 4.11

KEY CONCEPTS - simulating a random phenomenon

a) You will need to use your computer software to simulate the 100 trials. As the problem suggests, the key phrase to look for in your software is "Bernoulli trials." Your software may allow you to actually generate the words "Hit" and "Miss" or perhaps the letters H and M. More likely, your software will allow you to generate only the numbers 0 and 1. In this case, count a 0 as a miss and a 1 as a hit.

After you do this, the computer can be used to calculate the proportion of hits.

Proportion of hits =

For most students, the proportion of hits will be within 0.05 or 0.10 of the true probability of 0.5.

b) You need to go through your sequence to determine the longest string of hits or misses.

Longest run of shots hit =   Longest run of shots missed =

COMPLETE SOLUTIONS

Exercise 4.1

We desire information about the entire carload lot of ball bearings. The inspector chooses 100 bearings from the lot in order to decide whether to accept or reject the entire lot. In this problem, the entire carload lot is the population and the 100 bearings the sample. Thus

> 2.5003 cm. = a parameter since it describes the entire carload lot bearings (the population).
>
> 2.5009 cm. = a statistic since it describes the sample of 100 bearings.

Exercise 4.7

a) We obtained the following numbers of tosses until the first head for our 50 trials.

2	5	2	2	4	1	1	3	5	2
1	3	1	1	1	2	1	1	3	1
3	1	1	3	1	2	2	1	1	1
1	1	9	2	2	2	1	7	1	1
3	1	1	1	2	1	1	2	1	1

Recall that the probability of an event is the proportion of times the event occurs in many repeated trials of a random phenomenon. Therefore, to estimate the probability of a head on the first toss, we simply determine the proportion of cases in which we got a head on the first toss (which is the number of times a 1 appears in the above list). We see that 1 appears 27 times, so the proportion of times a head appeared on the first toss is 27/50 or 0.54. Thus our estimate of the probability of heads on the first toss is 27/50 = 0.54.

We would expect this probability to have value 1/2 or 0.5, since we would expect heads and tails to be equally likely when we flip a penny. The difference between 0.54 and 0.50 is due to sampling variablility.

b) To estimate the probability that the first head appears on an odd-numbered toss, we determine the proportion of odd numbers in our list. There are 37 odd numbers in the list, so the proportion of odd numbers is 37/50 = 0.74. Thus our estimate of the probability that the first head appears on an odd-numbered toss is 37/50 = 0.74.

Exercise 4.9

The event of interest is obtaining three of a kind in a five-card poker hand. The random phenomenon is the sequence of shuffling a deck of 52 cards and then dealing a five-card poker hand. A trial is any specific five-card poker hand obtained as the result of shuffling and dealing. If we incorporate this into the notion of probability, we might explain the statement that the probability of being dealt three of a kind in a five-card poker hand is 1/50 as follows.

If you repeat the process of shuffling a deck of 52 cards and then dealing a five-card poker hand many, many times, and you keep track of the proportion of hands that contain three of a kind, you expect this proportion to be 1/50.

Exercise 4.11

a) Here is our sequence of hits (H) and misses (M).

H	H	M	H	H	H	M	M	H	H	H	M	H	M	H
M	H	M	M	H	M	M	H	H	H	M	M	H	H	H
M	M	M	M	H	M	M	H	H	H	H	M	H	H	H
M	M	M	M	M	H	M	H	H	M	H	M	H	M	M
H	H	H	H	M	H	M	M	M	M	H	H	M	H	H
M	M	H	H	H	M	M	H	H	M	M	H	M	H	M
M	H	M	H	H	M	H	H	H	H					

proportion of hits = .54

b) You need to go through your sequence to determine the longest string of hits or misses. In our example,

Longest run of shots hit = 4 (occurred more than once)
Longest run of shots missed = 5

SECTION 4.2

OVERVIEW

The description of a random phenomenon begins with the **sample space, S**, which is the list of all possible outcomes. A set of outcomes is called an **event.** Once we have determined the sample space, a **probability model** tells us how to assign probabilities to the various events that can occur. There are four basic rules that probabilities must satisfy.

- Any probability is a number between 0 and 1. $P(A)$ means "the probability of A." If the probability is 0, the event will never occur. If the probability is 1, the event will always occur.

- All possible outcomes together must have probability 1.

- The probability that an event does not occur is 1 minus the probability that the event occurs. Using our notation, this can be written as

$$P(A) = 1 - P(\text{not } A)$$

- If two events have no outcomes in common, the probability that one or the other occurs is the sum of their individual probabilities. These events are **disjoint.** This is the addition rule for disjoint events, namely

$$P(A \text{ or } B) = P(A) + P(B)$$

In a sample space with a finite number of outcomes, probabilities are assigned to the individual outcomes and the probability of any event is the sum of the probabilities of the outcomes it contains. All outcomes must have a probability between 0 and 1; the sum of all probabilities must add up to 1.

Even a sample space with an infinite number of outcomes can have probabilities assigned to it. To do this we use a density curve. (Refer back to Section 1.3 to refresh your memory.) The area under a density curve must be equal to 1. Probabilities are assigned to events as areas under the density curve. The normal distribution is the most useful of the density curves. **Normal distributions are probability models**.

A **random variable** is a variable whose value is a numerical outcome of a random phenomenon. The **probability distribution** of a random variable tells us about the possible values of the random variable and how to assign probabilities to these values.

GUIDED SOLUTIONS

Exercise 4.15

KEY CONCEPTS - sample space

One of the main difficulties encountered when describing the sample space is finding some notation to express ideas formally. Following the text, our general format is $S = \{ \quad \}$, where a description of the outcomes in the sample space is included within the braces.

a) You want to express that any number between 0 and 24 is a possible outcome. So you would write $S = \{$all numbers between 0 and 24$\}$.

b) $S =$

c) $S =$

d) You may not want to put an upper bound on the amount, so you should allow for any number greater than 0.

$$S =$$

e) Remember that the rats can lose weight.

$$S =$$

Exercise 4.21

KEY CONCEPTS - applying the probability rules

a) Since these are the only blood types, what has to be true about the sum of the probabilities for the different types? Use this to find $P(AB)$.

b) What's true about the events O and B blood type? Which probability rule do we follow? (Don't be confused by the wording in the problem, which says "people with blood types O <u>and</u> B." In the context of this problem and the language of probability we are using, it really means O or B. There are no people with blood types O <u>and</u> B).

Exercise 4.31

KEY CONCEPTS - sample spaces for simple random sampling, probabilities of events

a) It's easy to make the list. $S = \{(\text{Abby, Deborah}), (\text{Abby, Sam}), \text{etc.}\}$. There is no need to include both (Abby, Deborah) and (Deborah, Abby) in your list, since both refer to the same two individuals.

b) How many outcomes are there in the sample space in part a? If they are equally likely, what is the probability of each?

c) How many outcomes in S include Tonya? When the outcomes are equally likely, the probability of the event is just

$$\frac{\text{number of outcomes in the event}}{\text{number of outcomes in } S}$$

d) How many outcomes in S include neither of the two men?

Exercise 4.35

KEY CONCEPTS - random variables, probability distributions

a) Write the event in terms of a probability about the random variable X. While you can figure out the answer without doing this, it's good practice to start using the notation for random variables.

b) Check to see that all the probabilities are numbers between 0 and 1 and that they sum to 1.

c) Add up the probabilities for values of $X \leq 3$.

d) Add up the probabilities for values of $X < 3$. Make sure that you understand the difference between this answer and what was required for part c.

e) Events about X are typically of the form bigger than or bigger than or equal to some number, or less than or less than or equal to some number. Write the event in one of these forms.

Exercise 4.37

KEY CONCEPTS - random numbers, density curve

a) The total area underneath any density curve is 1. Since the density curve has constant height over the range 0 to 2, what must the height be to make the area under this curve equal to 1? Note that since the area under the density curve is a rectangular region, the formula for the area of a rectangle will be useful.

Now draw a graph of the density curve.

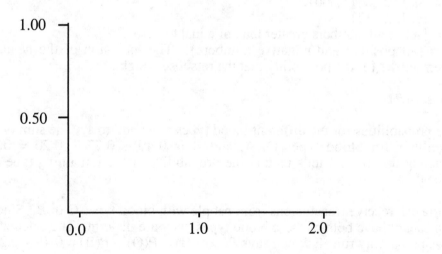

b) $P(Y \leq 1)$ is the area under the density curve below 1. You may find it helpful to sketch this region on your graph in part a. This is a rectangular region, so use the formula for the area of a rectangle to compute this area.

---

c) $P(0.5 \leq Y \leq 1.3)$ is the area under the density curve between 0.5 and 1.3. Compute this area. If you have trouble visualizing the region, sketch it on your part a graph.

d) Try this one without any hints.

COMPLETE SOLUTIONS

Exercise 4.15

a) $S = \{$all numbers between 0 and 24$\}$.
b) $S = \{0, 1, 2, ..., 11,000\}$.
c) $S = \{0, 1, ..., 12\}$.
d) $S = \{$set of all numbers greater than or equal to zero$\}$.
e) $S = \{$all positive and negative numbers$\}$. The inclusion of the negative numbers allows for the possibility that the rats lose weight.

Exercise 4.21

a) The probabilities for the different blood types must add to 1. The sum of the probabilities for blood types O, A, and B is $0.49 + 0.27 + 0.20 = 0.96$. Subtracting this from 1 tells us that the probability of the remaining type AB must be 0.04.

b) Maria can receive transfusions from people with blood types O or B. Since a person cannot have both of these blood types, they are disjoint. The calculation follows probability rule 4, which says $P(O \text{ or } B) = P(O) + P(B) = 0.49 + 0.20 = 0.69$.

Exercise 4.31

a) $S = \{$(Abby, Deborah), (Abby, Sam), (Abby, Tonya), (Abby, Roberto), (Deborah, Sam), (Deborah, Tonya), (Deborah, Roberto), (Sam, Tonya), (Sam, Roberto), (Tonya, Roberto)$\}$.

b) There are 10 possible outcomes. Since they are equally likely, each has probability 0.10.

c) Tonya is in four of the outcomes, so her chance of attending the conference in Paris is $4/10 = 0.4$. (Note that each person has the same probability of going. Remember from Chapter 2 this is a property of an SRS).

d) The chosen group must contain two women, (Abby, Deborah), (Abby, Tonya), or (Deborah, Tonya). There are three possibilities, so the desired probability is $3/10 = 0.3$.

Exercise 4.35

a) $P(X = 5) = 0.01$

b) All the probabilities are between 0 and 1 and the sum of the probabilities is $0.48 + 0.38 + 0.08 + 0.05 + 0.01 = 1.00$.

c) For the probability that $X \leq 3$, add the probabilities corresponding to $X = 1, 2$, and 3 to give $P(X \leq 3) = 0.48 + 0.38 + 0.08 = 0.94$.

d) The probability for $X < 3$ includes only values of X that are strictly below 3 that correspond to $X = 1$ and 2. So $P(X < 3) = 0.48 + 0.38 = 0.86$.

e) The son reaches one of the two higher classes if X is equal to 4 or 5. In terms of a single expression, this event is $X \geq 4$. $P(X \geq 4) = 0.05 + 0.01 = 0.06$. (Notice that this is the complement of the event in part c, which is why the two probabilities sum to 1.)

Exercise 4.37

a) The density curve will be a rectangle with base covering the region 0 to 2. The base has length 2. The area of a rectangle is base x height, so the area of the density curve is 2 x height. This area must equal 1, hence the height must be 0.5.

Here is a graph of the density curve.

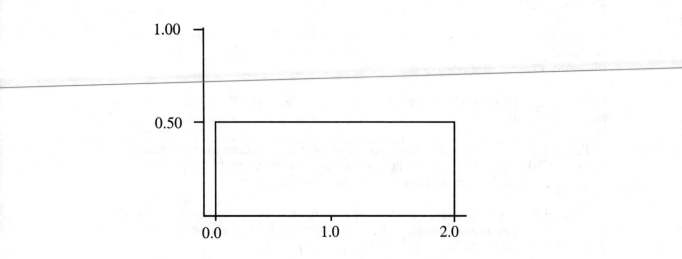

b) Here is a graph of the desired region.

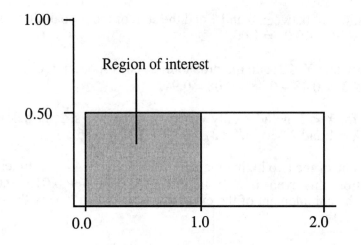

The region of interest is a rectangle with base of length 1.0 and height 0.5. Thus the area of this rectangle is $1 \times 0.5 = 0.5$, so

$$P(Y \leq 1) = 0.5$$

c) Here is a graph of the desired region.

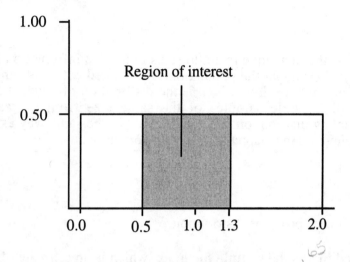

The region of interest this time is a rectangle with base of length 1.3 - 0.5 = 0.8 and height 0.5. Thus the area of this rectangle is 0.8 x 0.5 = 0.4, so

$$P(0.5 \leq Y \leq 1.3) = \underline{0.4}$$

d) Here is a graph of the desired region.

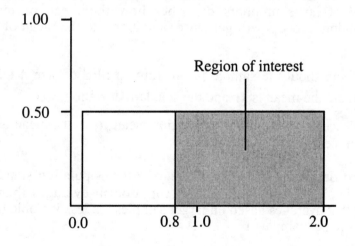

The region of interest this time is a rectangle with base of length 2.0 - 0.8 = 1.2 and height 0.5. Thus the area of this rectangle is 1.2 x 0.5 = 0.6, so

$$P(Y \geq 0.8) = 0.6$$

SECTION 4.3

OVERVIEW

Statistical inference is the technique that allows us to use the information in a sample to draw conclusions about the population. Associated with any sample statistic is its **sampling distribution,** which is the distribution of values taken by the statistic in all possible samples of the same size from the same population. The sampling distribution can be described in the same way as the distributions we encountered in Chapter 1. Three important features are

- measure of center

- measure of spread

- description of the shape of the distribution

The statistic discussed first is the **sample mean**, $\bar{x}$, which is an estimate of the **population mean** μ. $\bar{x}$ has a number of convenient properties when taken from an SRS that allow one to make a variety of inferences about μ. As with all sampling distributions, we need to know the mean, standard deviation, and the shape of the distribution. In fact, we know from the **central limit theorem** that the shape of the distribution of the sample mean is very close to normal when simple random sampling is used and sample sizes are large. The **law of large numbers** also tells us more about the behavior of $\bar{x}$ as the sample size increases. The law of large numbers describes how the mean of many observations of a random process will get closer and closer to the mean of the population.

Here are the basic facts about the sample mean from an SRS of size n taken from a population where the mean is μ and the standard deviation is σ:

- $\bar{x}$ is an **unbiased estimate** of the population mean μ, so the mean of $\bar{x}$ is the population mean μ.

- The standard deviation is $\sigma/\sqrt{n}$, where σ is the population standard deviation. (This shows that there is less variation in averages than in individuals and that averages based on larger samples are less variable than those based on smaller samples.)

- The shape of the distribution of the sample mean depends on the shape of the population distribution. If the population was normal, $N(\mu, \sigma)$, then the sample mean has normal distribution $N(\mu, \sigma/\sqrt{n})$. Also, for a large sample the central limit theorem tells us that the sample mean has *approximately* a normal distribution $N(\mu, \sigma/\sqrt{n})$.

A statistic is called **unbiased** if the sampling distribution of the statistic is centered (has its mean) at the value of the population parameter. This means that the statistic tends to neither overestimate nor underestimate the parameter.

GUIDED SOLUTIONS

Exercise 4.39

KEY CONCEPTS - law of large numbers

Review the statement of the law of large numbers in the text. In this problem, μ is 60 cents. $\bar{x}$ would be the amount Joe receives as payoff per bet on average after many years (i.e., the total amount paid to Joe divided by the number of bets placed). What does the law of large numbers say about $\bar{x}$? Since Joe must pay \$1.00 for each bet, what can you say about Joe's average net winnings per bet (average winnings per bet minus the cost of the bet)?

Exercise 4.41

KEY CONCEPTS - standard deviation of the sampling distribution of $\bar{x}$

a) The key formula is that if $\bar{x}$ is the mean of an SRS of size n drawn from a large population with mean μ and standard deviation σ, then the standard deviation of the sampling distribution of $\bar{x}$ is $\dfrac{\sigma}{\sqrt{n}}$. To apply this here, identify σ and n and then complete the following:

$$\text{Standard deviation of Juan's mean result} = \frac{\sigma}{\sqrt{n}} =$$

b) What value would n have to be so that $\dfrac{\sigma}{\sqrt{n}}$ is 5?

Write out your explanation of the advantage of reporting the average of several measurements rather than the result of a single measurement. Remember, don't use technical language.

Exercise 4.47

KEY CONCEPTS - sampling distributions, the central limit theorem

a) You may wish to refer to Section 1.1 to refresh your memory about how to make a histogram. You may use the axes here to assist you.

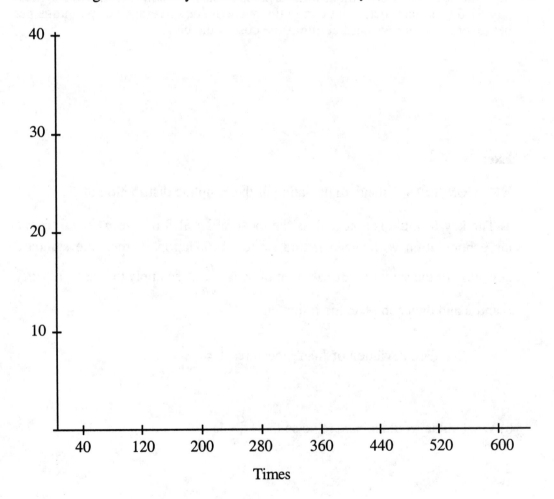

Times

b) You might find it helpful to read Exercise 1.80. Mark μ on the horizontal axis of the histogram in part a.

c) The 72 survival times are reproduced here to make you. Label these values from 01 to 72 and proceed to choose an SRS of size 12. Refer to Chapter 3 if you need to review how to choose an SRS.

43	45	53	56	56	57	58	66	67
73	74	79	80	80	81	81	81	82
83	83	84	88	89	91	91	92	92
97	99	99	100	100	101	102	102	102
103	104	107	108	109	113	114	118	121
123	126	128	137	138	139	144	145	147
156	162	174	178	179	184	191	198	211
214	243	249	329	380	403	511	522	598

Line in Table B used to select SRS =

SRS of size 12 =

Sample mean $\bar{x}$ =

Remember to mark $\bar{x}$ on your histogram in part a.

d) Write the results of your next four SRSs.

SRS 2 =

$\bar{x}$ =

SRS 3 =

$\bar{x} =$

SRS 4 =

$\bar{x} =$

SRS 5 =

$\bar{x} =$

Remember to mark these four new values of $\bar{x}$ on your histogram in part a.

Would you be surprised if all five $\bar{x}$'s fell on the same side of μ? Why?

e) Where would you expect the center of this sampling distribution to lie?

Exercise 4.51

KEY CONCEPTS - the sampling distribution of the sample mean, normal probability calculations

a) You may wish to review normal probability calculations in Section 1.3 to refresh your memory. What is the z score of 295?

z score =

Shade in the desired area in the graph below.

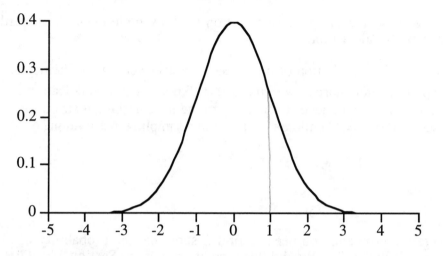

Now use Table A to find this area.

Area =

b) Now we are interested in probabilities concerning the mean of the contents of six bottles. What does the Central Limit Theorem have to say about the sampling distribution of the mean based on samples of size 6? What, therefore, is the appropriate sampling distribution for the probability requested?

Sampling distribution =

Based on this sampling distribution, what is the z score of 295 here?

z score =

Draw a graph of the desired area under a standard normal curve and use Table A to find this area.

Exercise 4.55

KEY CONCEPTS - the sampling distribution of the sample mean, backward normal probability calculations

We are told that the distribution of individual scores at Southwark Elementary School is approximately normal with mean $\mu = 13.6$ and standard deviation $\sigma = 3.1$. To find L, Mr. Lavin needs to first determine the sampling distribution of the mean score $\bar{x}$ of $n = 22$ children. What is this sampling distribution?

To complete the problem, you need to find L such that the probability of $\bar{x}$ being below L is only 0.05. We did this type of problem in Section 1.3. First, refer to Table A to find the value z such that the area to the left of z under a standard normal curve is 0.05. What is this value?

This value z is the z score of L. This means $z = (L - \mu_{\bar{x}})/\sigma_{\bar{x}}$, where $\mu_{\bar{x}}$ and $\sigma_{\bar{x}}$ are the mean and standard deviation, respectively, of the sampling distribution of $\bar{x}$. Solve this equation for L.

COMPLETE SOLUTIONS

Exercise 4.39

The law of large numbers tells us that after many years, Joe should receive a payoff of about 60 cents per bet, on average. However it costs Joe $1.00 to make a bet. Hence his net average winnings after many years should be $0.60 - $1.00 = -$0.40. In other words, in the long run, Joe will lose 40 cents per bet on average.

Exercise 4.41

a) In this problem $\sigma = 10$ and $n = 3$. Thus

$$\text{Standard deviation of Juan's mean result} = \frac{\sigma}{\sqrt{n}} = \frac{10}{\sqrt{3}} = 5.77$$

b) We want

$$5 = \frac{\sigma}{\sqrt{n}} = \frac{10}{\sqrt{n}}$$

Solving this equation for $\sqrt{n}$, we must have

$$\sqrt{n} = 10/5 = 2$$

Squaring both sides yields $n = 4$.

Averages of several measurements are more likely to be closer to the true value of the quantity being measured than a single measurement. The magnitude of chance deviations or errors are smaller for averages than for individual observation.

Exercise 4.47

a) Here is a histogram of the 72 survival times.

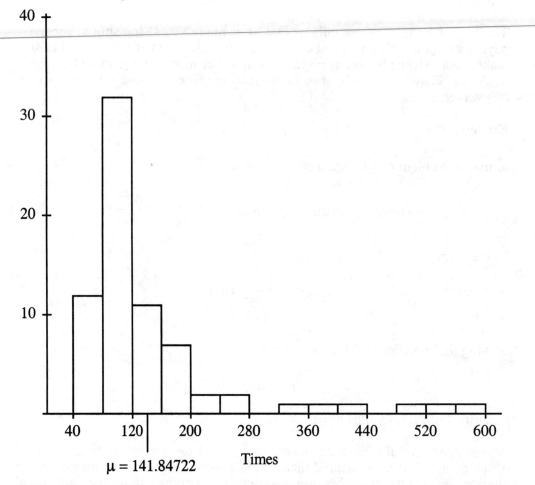

$\mu = 141.84722$

b) Exercise 1.80 gives the mean as 141.84722. Of course, you can also compute the mean directly from the 72 survival times. See the histogram in part a which has μ marked on the horizontal axis.

c) We label times from left to right. Labels are in parentheses.

(01) 43	(02) 45	(03) 53	(04) 56	(05) 56	(06) 57	(07) 58	(08) 66	(09) 67
(10) 73	(11) 74	(12) 79	(13) 80	(14) 80	(15) 81	(16) 81	(17) 81	(18) 82
(19) 83	(20) 83	(21) 84	(22) 88	(23) 89	(24) 91	(25) 91	(26) 92	(27) 92
(28) 97	(29) 99	(30) 99	(31) 100	(32) 100	(33) 101	(34) 102	(35) 102	(36) 102
(37) 103	(38) 104	(39) 107	(40) 108	(41) 109	(42) 113	(43) 114	(44) 118	(45) 121
(46) 123	(47) 126	(48) 128	(49) 137	(50) 138	(51) 139	(52) 144	(53) 145	(54) 147

(55) 156   (56) 162   (57) 174   (58) 178   (59) 179   (60) 184   (61) 191 (62) 198 (63) 211

(64) 214   (65) 243   (66) 249   (67) 329   (68) 380   (69) 403   (70) 511 (71) 522 (72) 598

The line in Table B used to select SRS is 188. The SRS of size 12 (labels) is 37, 08, 69, 58, 33, 55, 03, 26, 25, 17, 36, 29. Note that to obtain the last number we had to continue on line 189.

The actual times corresponding to these labels are 103, 66, 403, 178, 101, 156, 53, 92, 91, 81, 102, 99, and if we compute their mean we get

$$\text{Sample mean } \ \bar{x} \ = 127.08$$

Following is the histogram with $\bar{x}$ marked on the horizontal axis.

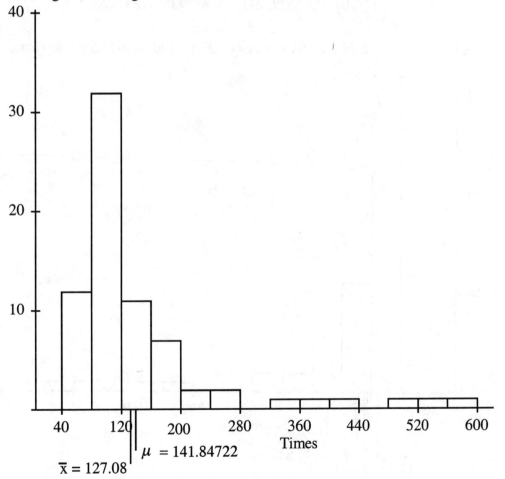

d) Here are the next four SRSs and the corresponding values of $\bar{x}$. The samples were selected beginning on line 189 where we left off in part c. We continued on through line 192. For brevity, we list only the times corresponding to the labels selected followed by the mean of these times.

SRS 2 = 403, 162, 511, 92, 145, 81, 107, 126, 179, 99, 118, 128
$\quad$ $\bar{x}$ = 179.25

SRS 3 = 179, 88, 108, 81, 118, 102, 101, 380, 92, 56, 121, 53
$\quad$ $\bar{x}$ = 123.25

SRS 4 = 114, 108, 45, 137, 89, 101, 83, 162, 88, 179, 109, 128
$\quad$ $\bar{x}$ = 111.92

SRS 5 = 243, 107, 100, 329, 211, 45, 88, 73, 57, 83, 138, 118
$\quad$ $\bar{x}$ = 132.67

Here is the histogram with all five values of $\bar{x}$ marked on the horizontal axis.

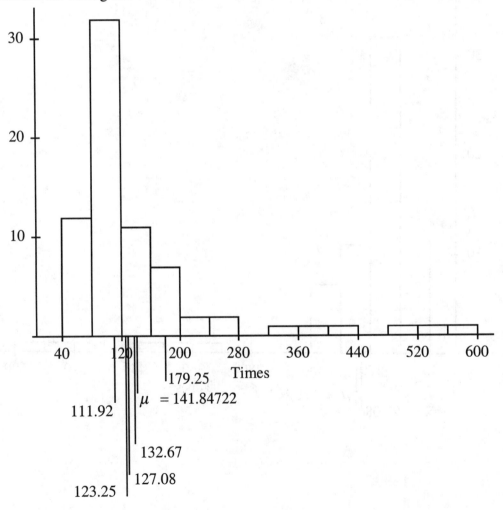

We would be somewhat surprised if all five $\bar{x}$'s fell on the same side of μ. The central limit theorem suggests that the sampling distribution of the $\bar{x}$'s should have a mean equal to μ and that the sampling distribution might be expected to look very roughly bell-shaped and symmetric. This would suggest that the probability that $\bar{x}$ would be above μ would be about 1/2, the same for the probability that $\bar{x}$ would be below μ. Thus getting all five $\bar{x}$'s on the same side of μ would be like flipping a coin five times and getting either all heads or all tails, a relatively uncommon occurrence.

e) As discussed in part d, the central limit theorem suggests that the sampling distribution of the $\bar{x}$'s should have a mean equal to μ (notice that $\bar{x}$ is an unbiased estimator of μ) and that the sampling distribution might be expected to look very roughly bell-shaped and symmetric. This suggests that μ would be a reasonable measure of the center of this distribution.

Exercise 4.51

a) The probability of interest is displayed as the shaded are in the graph.

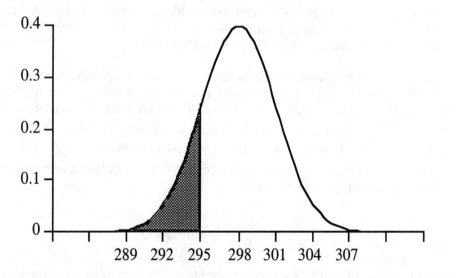

To find this area, we must compute th z-score of 295. Since the contents follow a normal distribution with mean $\mu = 298$ and standard deviation $\sigma = 3$,

$$z \text{ score} = \frac{x - \mu}{\sigma} = \frac{295 - 298}{3} = \frac{-3}{3} = -1.$$

In terms of areas under the standard normal, the area of interest is

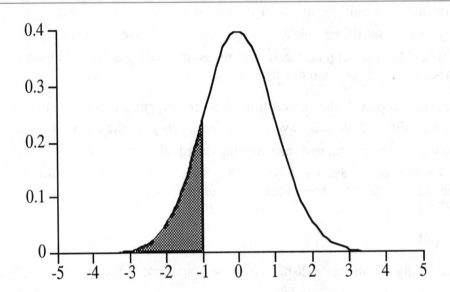

which can be read directly from Table A. From Table A we find that the area to the left of -1 under a standard normal curve is 0.1587. Therefore the probability that an individual bottle contains less than 295 ml is 0.1587.

b) For questions about probabilities associated with the mean contents of the bottles in a six-pack, we need to determine the sampling distribution of the mean of six bottles. The contents of individual bottles vary according to a normal distribution with mean μ = 298 and standard deviation σ = 3. According to the central limit theorem, this sampling distribution for the mean of the contents of a six-pack is normal with the mean μ = 298 (the same as for individual bottles) and standard deviation

$$\frac{\sigma}{\sqrt{n}} = \frac{3}{\sqrt{6}} = 1.22$$

To compute the probability that the mean contents of the bottles in a six-pack is less than 295, we again must find the z score of 295. This time,

$$z \text{ score} = \frac{295 - 298}{1.22} = \frac{-3}{1.22} = -2.46.$$

In terms of areas under the standard normal, the area of interest is now the shaded area in the following graph.

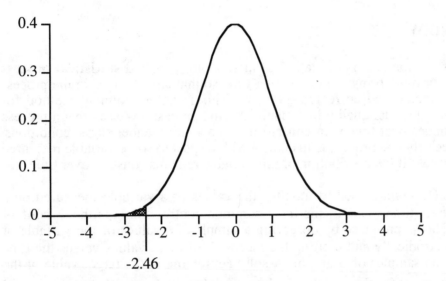

-2.46

This area can be read directly from Table A. From Table A we find that the area to the left of -2.46 under a standard normal curve is 0.0069. Therefore the probability that the mean contents of the bottles in a six-pack is less than 295 ml is 0.0069.

EXERCISE 4.55

The sampling distribution of the mean score $\bar{x}$ of 22 children is approximately normal with mean

$$\mu_{\bar{x}} = \mu = 13.6$$

and standard deviation

$$\sigma_{\bar{x}} = \frac{\sigma}{\sqrt{n}} = \frac{3.1}{\sqrt{22}} = 0.66$$

We note that from Table A the value of z such that the area to the left of it under a standard normal curve is 0.05 is $z = -1.65$. Thus

$$-1.65 = (L - 13.6)/0.66$$

Solving for L gives

$$L = (-1.65)(0.66) + 13.6 = 12.51$$

SECTION 4.4 (optional section from the CD)

OVERVIEW

This section examines statistical methodology, called **statistical process control**, for monitoring a process over time so that any changes in the process can be detected and corrected quickly. This is an economical method for maintaining product quality in a manufacturing process. We say that a process that continues over time is in **control** if it is operating under stable conditions. More precisely, the process is in control with respect to some variable measured on the process if the distribution of this variable remains constant over time.

Control charts are used to monitor the values of a variable measured on a process over time. One of the most common control charts is the $\bar{x}$ **control chart**. This is produced by observing a sample of n values of the variable of interest periodically and plotting the means $\bar{x}$ of these values versus the time order of the samples on a graph. A solid **centerline** at the target value of the process mean μ for the variable is drawn on the graph, as are dashed **control limits** at

$$\mu \pm 3\frac{\sigma}{\sqrt{n}}$$

where σ is the process standard deviation of the variable. This chart helps us decide if the process is in control with mean μ and standard deviation σ. The probability that the next point (value of $\bar{x}$) on such a chart lies outside the control limits is about 0.003 if the process is in control. Such a point would be evidence that the process is **out of control**, that is, that the distribution of the process has changed for some reason. When a process is deemed out of control, a cause for the change in the process should be sought.

In addition to a point plotting outside the control limits, there are other signals that suggest that the process is out of control. One is a **run** of nine consecutive points on the same side of the centerline. Whatever signal is used, we can compute the probability that a process that is in control will give the signal. If this probability is small, we can regard the signal as evidence that the process is out of control.

In the real world we seldom know the values of μ and σ. When these parameters are not known, we construct the $\bar{x}$ control chart as follows. Compute $\bar{\bar{x}}$, the average of the $\bar{x}$ values computed for each of our samples of size n. Also compute the standard deviations s for each of the samples of size n and from these compute $\bar{s}$, the mean of these s values. The $\bar{x}$ control chart is a plot of the $\bar{x}$'s against time with center line $\bar{\bar{x}}$ and control limits $\bar{\bar{x}} \pm A\bar{s}$. The constant A depends on n and can be found in Table 4.3.

With such an $\bar{x}$ control chart, it is customary to also construct an **s chart**. The s chart plots the standard deviations s of a certain characteristic from samples of size n versus the time order of the samples. The s chart uses a center line of $\bar{s}$, the mean of the s values plotted, and control limits $\bar{s} \pm B\bar{s}$, where B depends on n and can be found in Table 4.3.

GUIDED SOLUTIONS

Exercise 4.72

KEY CONCEPTS - $\bar{x}$ control chart with μ and σ known

When the process is properly adjusted, you need to identify

$\mu =$

$\sigma =$

$n =$ sample size $=$

From this information, determine

Center line $=$

Control limits $=$

Refer to the section overview for the equations of the center line and control limits for $\bar{x}$ charts when μ and σ are known.

Now graph the $\bar{x}$ control chart.

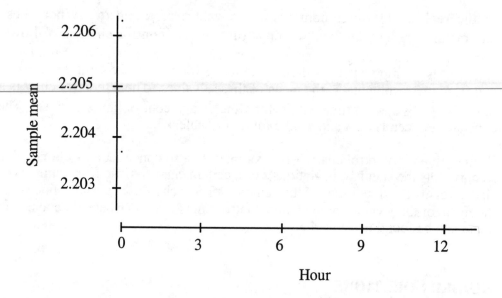

Apply the one point out and run of nine rules to your chart to assess the control of the process.

Exercise 4.75

KEY CONCEPTS - $\bar{x}$ and s control chart with μ and σ unknown

a) To construct an s chart, compute

$\bar{s}$ = the mean of the 15 values of s =

n =

Now compute

Center line =

Control limits =

Refer to the section overview for the formulas for the center line and control limits for s charts. Remember that if the lower control limit is negative, we ignore it.

Now graph the s control chart.

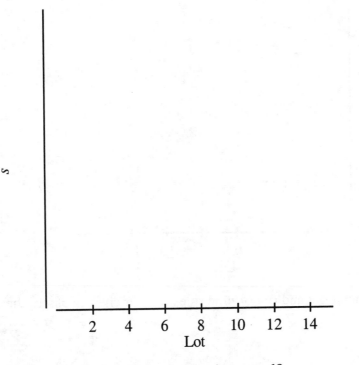

Is the short-term variation of the process in control?

b) To construct an $\bar{x}$ control chart, compute

$\bar{\bar{x}}$ = the average of the values of $\bar{x}$ =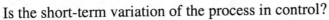

Now compute

Center line =

Control limits =

Refer to the section overview for the appropriate formulas for $\bar{x}$ control charts when μ and σ are not known.

Now graph the control chart.

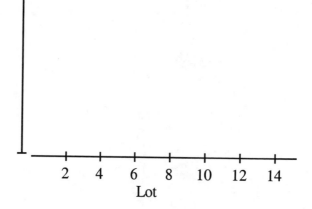

Apply the one point out and run of nine rules to your chart to assess the control of the process.

Exercise 4.76

KEY CONCEPTS - normal probabilities, $\bar{x}$ control charts

a) We want to know the probability that the line width, which varies according to a normal distribution with mean $\mu = 2.829$ mm and standard deviation $\sigma = 0.1516$ mm, takes on a value below $3.0 - .2$ mm $= 2.8$ mm or above $3.0 + .2$ mm $= 3.2$ mm. This probability is calculated using normal probability calculations such as you did in Section 1.3 or in Section 4.3.

b) You should be able to do this on your own without hints.

COMPLETE SOLUTIONS

Exercise 4.72

When the process is properly adjusted, the mean is $\mu = 2.2050$ cm with standard deviation $\sigma = .0010$ cm. For means based on samples of size 5, the center line is the horizontal line at $\mu = 2.2050$ and the control limits are horizontal lines at

$$\mu \pm 3\sigma/\sqrt{n} = 2.2050 \pm 3(.0010)/\sqrt{5} = 2.2050 \pm .0013$$

or at 2.2037 and 2.2063. The $\bar{x}$ control chart is thus

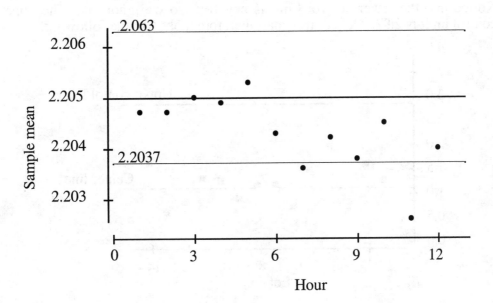

Using the one point out signal, we are alerted at samples 7 and 11 that the process appears out of control. Action should have been taken at sample 7. Using the run of nine signal, the process never appears out of control (although the last 8 points are all on the same side of the center line).

Exercise 4.75

a) To construct an s chart, we must compute $\bar{s}$, the mean of the 15 values of s. We find

$$\bar{s} = 1.11667$$

Thus the center line of the s chart is

$$\text{Center line of } s \text{ chart} = \bar{s} = 1.11667$$

To compute the control limits $\bar{s} \pm B\bar{s}$, we need to determine the constant B. Each sample consists of $n = 3$ insulators from a lot. According to Table 4.3,

$$B = 1.568$$

and hence the s chart has

$$\text{Control limits} = \bar{s} \pm B\bar{s} = 1.11667 \pm 1.568 \times 1.11667 = 1.11667 \pm 1.75094$$

Notice that the lower control limit is negative, so we ignore it. The upper control limit is 2.87. Our control chart therefore looks like the following.

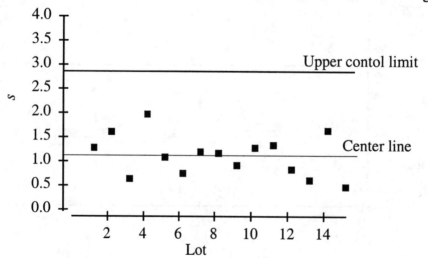

We see that the short-term variation of the process is in control.

b) To construct an $\bar{x}$ control chart we must compute $\bar{\bar{x}}$, the average of the values of $\bar{x}$. We find

$$\bar{\bar{x}} = 11.8307$$

Thus

Center line of $\bar{x}$ control chart $= \bar{\bar{x}} = 11.8307$

The control limits are $\bar{\bar{x}} \pm A\bar{s}$. From part a we know that $\bar{s} = 1.11667$. From Table 4.3 we find that $A = 1.954$. Therefore

Control limits $= \bar{\bar{x}} \pm A\bar{s} = 11.8307 \pm 1.945 \times 1.11667 = 11.8307 \pm 2.1719$.

The lower control limit is 9.66 and the upper control limit is 14.00. Following is the $\bar{x}$ control chart.

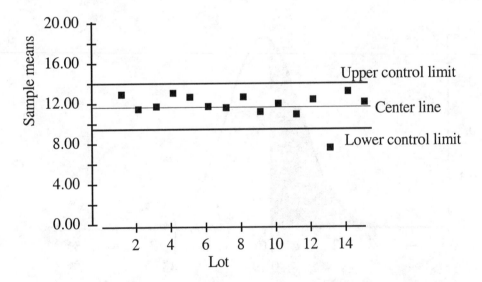

These data do not have a run of nine, but one point (the one corresponding to lot 13) is outside the lower control limit. Since this is an unusually low value, we would want to check whether there is anything unusual about lot 13. For example, was the ceramic in this lot improperly formed or baked? The process is again within control limits for lots 14 and 15, so one might suspect that the problem lies solely with lot 13, not with the process after lot 13.

Exercise 4.76

a) We want to know the probability that the line width X, which varies according to a normal distribution with mean $\mu = 2.829$ mm and standard deviation $\sigma = 0.1516$ mm, takes on a value below $3.0 - .2$ mm $= 2.8$ mm or above $3.0 + .2$ mm $= 3.2$ mm. This probability is

$$P(X \le 2.8) + P(X \ge 3.2) = P\left(\frac{X - 2.829}{0.1516} \le \frac{2.8 - 2.829}{0.1516}\right) +$$

$$P\left(\frac{X - 2.829}{0.1516} \ge \frac{3.2 - 2.829}{0.1516}\right)$$

$$= P(Z \le -0.19) + P(Z \ge 2.45)$$
$$= .4247 + .0071$$
$$= .4318$$

This probability is displayed as the shaded portion of the normal curve in the following graph.

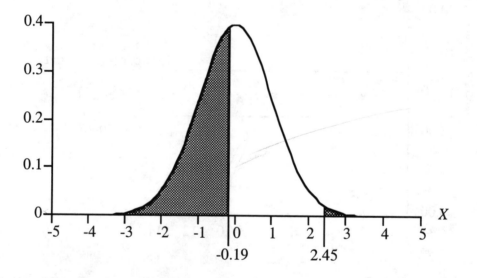

The percent of chips outside the acceptable range is therefore 43.18%.

b) Assuming that the target value of 3.0 is the value of the center line, μ, the control limits for an $\bar{x}$ chart for line width if samples of size 5 are used are

$$\mu \pm 3\sigma/\sqrt{n} = 3.0 \pm 3(.1516)/\sqrt{5} = 3.0 \pm .2034$$

Thus the lower control limit is the horizontal line at 2.7966 and the upper control limit is the horizontal line at 3.2034.

CHAPTER 5

PROBABILITY THEORY

SECTION 5.1

OVERVIEW

Chapter 4 covers the four basic rules of probability. In this chapter we will use the basic rules, combined with some additional ones, to calculate probabilities of more complex events.

Recall the addition rule for disjoint events,

$$P(A \text{ or } B) = P(A) + P(B)$$

Events are **independent** if knowledge that one event has occurred does not alter the probability that the second event occurs. The mathematical definition of independence leads to the **multiplication rule** for independent events. If A and B are independent, then

$$P(A \text{ and } B) = P(A)P(B)$$

In any particular problem we can use this definition to check if two events are independent by seeing if the probabilities multiply according to the definition. However, most of the time, independence is assumed as part of the probability model. The four basic rules, plus the multiplication rule, allow us to compute the probabilities of events in many random phenomena.

Many students confuse independent and disjoint events. Remember, disjoint events have no outcomes in common; when two events are disjoint, you can compute $P(A \text{ or } B) = P(A) + P(B)$. The probability being computed is that one *or* the other event occurs. Disjoint events cannot be independent since once we know that A has occurred, then the probability of B occurring becomes 0 (B cannot have occurred as well - this is the meaning of disjoint). The multiplication rule can be used to compute the probability that two events occur *simultaneously*, $P(A \text{ and } B) = P(A)P(B)$, in the special case of independence. This rule can be extended to more than two events as long as all are independent of the others.

To calculate $P(A \text{ or } B)$ for events that are not disjoint, we use the **general addition rule:**

$$P(A \text{ or } B) = P(A) + P(B) - P(A \text{ and } B)$$

GUIDED SOLUTIONS

Exercise 5.3

KEY CONCEPTS - independence, multiplication rule

The key concept that must be properly understood to answer the questions raised in this problem is the notion of independence. Events A and B are independent if knowledge that A has occurred does not alter our assessment of the probability that B will occur.

What is required to use the multiplication rule to find the probability that A and B occur? Do you think the multiplication rule applies here?

Exercise 5.5

KEY CONCEPTS - independence, multiplication rule

Suppose A is the event that test A is positive if steroids have been used and not A is the event that test A is not positive. Similarly, B is the event that test B is positive if steroids have been used and not B is the event that test B is not positive. Remember that if events A and B are independent, the event that A does not occur is also independent of B, and so on. Try to write the event that neither test is positive in terms of not A and not B and then use the multiplication rule.

Exercise 5.9

KEY CONCEPTS - independence, multiplication rule

The probability of winning the major battle is 0.6. What is the probability of winning all three small battles? How would you decide which strategy is best? Write the event winning all three small battles in terms of winning each of the small battles and use the independence of victories or defeats in the small battles.

P(winning all three small battles) =

Which strategy do you prefer and why?

Exercise 5.15

KEY CONCEPTS - independence, events, addition rule, multiplication rule

In many problems, one of the greatest difficulties is deciding on some sort of notation that will allow you to express the problem simply. In this case, the following notation for events should prove helpful. For the husband, you can use the events H_A, H_B, H_O and H_{AB} to denote the events that the husband has type A, type B, type O, and type AB, respectively. For the wife the notation for these events is W_A, W_B, W_O, and W_{AB}. With these notational conventions, you can express more complicated events simply and then apply the addition and multiplication rules learned in this section to compute the required probabilities.

a) The probability requested is $P(H_B \text{ or } H_O)$. What rule can be used to find this probability?

$P(H_B \text{ or } H_O) =$

b) The probability requested is $P(W_B \text{ and } H_A)$. Remember that blood types of married couples are independent, and the probabilities given for the blood types are valid for both men and women. What rule can be used to find the required probability?

$$P(W_B \text{ and } H_A) =$$

c) The probability requested is not the same as in part b, although some of the calculations are the same. In this part we are not told which spouse has type A blood and which spouse has type B blood. The probability requested is given. See if you can apply the addition and multiplication rules correctly to get to the answer.

$$P([W_B \text{ and } H_A] \text{ or } [H_B \text{ and } W_A]) =$$

d) First find the probability that neither person in a randomly chosen couple has type O blood. Let the event $H_{\text{not O}}$ and $W_{\text{not O}}$ correspond to the husband not having type O blood and the wife not having type O blood, respectively. First evaluate

$$P(H_{\text{not O}}) = \qquad P(W_{\text{not O}}) =$$

Now use the events $H_{\text{not O}}$ and $W_{\text{not O}}$ to evaluate

$$P(\text{neither has type O}) =$$

Now you are almost done. How does $P(\text{at least one has type O})$ relate to $P(\text{neither has type O})$?

COMPLETE SOLUTIONS

Exercise 5.3

Suppose the events college-educated and laborer or operator are independent. This would imply that knowing whether someone was college-educated would not change the probability that they were a laborer or operator. In terms of this problem, if the events were independent, 16% of the entire labor force would work as laborers or operators, and also 16% of the college-educated labor force would work as laborers or operators (knowledge of whether or not an individual has four years of college doesn't alter [increase or decrease] their chance of being a laborer or operator). The independence is what allows us to just multiply these probabilities together. The use of the formula $(0.27)(0.16) = 0.043$ to get the answer implies that 16% of the college-educated people are laborers or operators. However, we would guess that fewer than 16% of those with four years of college were laborers or operators, so multiplying the two probabilities together would not be the correct way to get the answer (the answer obtained by multiplying the two probabilities together is too large). These outcomes are not independent.

Exercise 5.5

Suppose not A is the event that test A is not positive if steroids have been used and not B is the event that test B is not positive if steroids have been used. Then the event that neither is positive is identical to the event not A and not B, or in terms of the probabilities,

$$P(\text{neither test is positive}) = P(\text{not } A \text{ and not } B)$$

Remembering that if events A and B are independent, the event that A does not occur is also independent of B, and so on, we know that the events not A and not B are independent. Using the multiplication rule for independent events then gives

$$P(\text{neither test is positive}) = P(\text{not } A \text{ and not } B) = P(\text{not } A)P(\text{not } B)$$

Finally, $P(\text{not } A) = 1 - P(A) = 0.1$ and $P(\text{not } B) = 1 - P(B) = 0.2$, so

$$P(\text{not } A \text{ and not } B) = P(\text{not } A)P(\text{not } B) = (0.1)(0.2) = 0.02$$

which is the final answer.

We have shown you the most direct way to do the problem. Another way to do this problem is to use the fact that

$$P(\text{neither test is positive}) = 1 - [P(\text{not } A \text{ and } B) + P(A \text{ and not } B) + P(A \text{ and } B)]$$

and then use the independence to compute the three individual probabilities on the right-hand side.

Exercise 5.9

Denote by W_1 the event that the general wins the first small battle, W_2 the event that the general wins the second small battle, and W_3 the event that the general wins the third small battle. Then

$$P(\text{winning all three small battles}) = P(W_1 \text{ and } W_2 \text{ and } W_3) = P(W_1)P(W_2)P(W_3)$$

since victories or defeats in the small battles are independent. We know that the probability of winning each small battle is 0.8, so

$$P(\text{winning all three small battles}) = P(W_1)P(W_2)P(W_3) = 0.8 \times 0.8 \times 0.8 = 0.512.$$

Since the general is more likely to win the major battle than to win all three small battles, his strategy should be to fight one major battle.

Exercise 5.15

a) The probability requested is $P(H_B \text{ or } H_O)$. The events H_B and H_O are disjoint (a person can't have two types of blood), so the addition rule for disjoint events applies:

$$P(H_B \text{ or } H_O) = P(H_B) + P(H_O) = 0.13 + 0.44 = 0.57$$

b) The probability requested is $P(W_B \text{ and } H_A)$. The events W_B and H_A are independent since the blood types of married couples are independent. (Note that these events are not disjoint. It's possible for both of them to occur together. When first learning about probability, students often confuse the ideas of independent and disjoint events). To evaluate the required probability we use the multiplication rule for independent events:

$$P(W_B \text{ and } H_A) = P(W_B)P(H_A) = 0.13 \times 0.37 = 0.0481$$

c) The probability requested is $P([W_B \text{ and } H_A] \text{ or } [H_B \text{ and } W_A])$, since, unlike part b, the event doesn't specify which member of the couple has type A blood and which member has type B blood. The events $[W_B \text{ and } H_A]$ and $[H_B \text{ and } W_A]$ are disjoint, so to start we can write

$$P([W_B \text{ and } H_A] \text{ or } [H_B \text{ and } W_A]) = P([W_B \text{ and } H_A]) + P([H_B \text{ and } W_A])$$

Each of the probabilities on the right-hand side can be computed as in part b, giving

$$P([W_B \text{ and } H_A]) + P([H_B \text{ and } W_A]) = 0.13 \times 0.37 + 0.13 \times 0.37 = 0.0962$$

d) First we evaluate $P(H_{\text{not O}}) = 1 - P(H_O) = 1 - 0.44 = 0.56$. Similarly, $P(W_{\text{not O}}) = 0.56$. Now,

$$P(\text{neither has type O}) = P(H_{\text{not O}} \text{ and } W_{\text{not O}}) = P(H_{\text{not O}})P(W_{\text{not O}})$$

where we have used the multiplication rule since $H_{\text{not O}}$ and $W_{\text{not O}}$ are independent. Both the probabilities on the right are equal to 0.56 so we have

$$P(\text{neither has type O}) = P(H_{\text{not O}})P(W_{\text{not O}}) = 0.56 \times 0.56 = 0.3136$$

Finally,

P(at least one has type O) = 1 - P(neither has type O) = 1 - 0.3136 = 0.6864

Although this is the most direct way to do the problem, there are other ways to set it up and arrive at the answer. If you set it up differently and got a different answer, than you probably omitted some cases.

SECTION 5.2

OVERVIEW

One of the most common situations giving rise to a **count X** is the **binomial setting**. The binomial setting consists of four assumptions about how the count was produced:

- The number n of observations is fixed.
- The n observations are all independent.
- Each observation falls into one of two categories called "success" and "failure."
- The probability of success p is the same for each observation.

When these assumptions are satisfied, the number of successes, X, has a **binomial distribution** with n trials and success probability p. For smaller values of n, the probabilities for X can be found easily using statistical software or the exact **binomial probability formula**. The formula is given by

$$P(X = k) = \binom{n}{k} p^k (1-p)^{n-k}$$

where $k = 0, 1, 2, ..., n$, and $\binom{n}{k} = \dfrac{n!}{k!(n-k)!}$ is called the **binomial coefficient**.

When the population is much larger than the sample, a count X of successes in an SRS of size n has approximately the binomial distribution with n equal to the sample size and p equal to the proportion of successes in the population.

The mean of a binomial random variable X is

$$\mu = np$$

and the standard deviation is

$$\sigma = \sqrt{np(1-p)}$$

When n is large, the count X is approximately $N(np, \sqrt{np(1-p)})$. This approximation should work well when $np \geq 10$ and $n(1-p) \geq 10$.

GUIDED SOLUTIONS

Exercise 5.19

KEY CONCEPTS - binomial setting

Four assumptions need to be satisfied to ensure that the count X has a binomial distribution. The number of observations or trials is fixed in advance, each trial results in one of two outcomes, the trials are independent, and the probability of success is the same from trial to trial. In this problem, think about how many observations or trials there are going to be.

Exercise 5.23

KEY CONCEPTS - binomial probabilities

X is the number of players among the 20 who graduate. According to the university's claim, X should have the binomial distribution with $n = 20$ and $p = 0.8$. You need to find the probability that exactly 11 out of 20 players graduate, or evaluate $P(X = 11)$. The exact binomial probability formula is

$$P(X = k) = \binom{n}{k} p^k (1-p)^{n-k}$$

where $k = 0, 1, 2, ..., n$ and $\binom{n}{k}$ is the binomial coefficient. Plug the appropriate values of n, k, and p in the formula to arrive at the answer.

$$P(X = 11) =$$

Exercise 5.29

KEY CONCEPTS - mean and standard deviation of a binomial count, normal approximation to the binomial distribution

a) When X is known to have a binomial distribution, you can use the formulas that express the mean and standard deviation of X in terms of n and p.

Mean =

Standard deviation =

b) This part can be done using the normal approximation for a count. First check to see that np and $n(1 - p)$ are both greater than or equal to 10. Then use the mean and standard deviation from part a and the normal approximation to evaluate $P(X \leq 170)$.

$np =$ and $n(1 - p) =$

Can the approximation be used?

$P(X \leq 170) =$

Exercise 5.33

KEY CONCEPTS - binomial probabilities, mean and standard deviation of a binomial count

a) A count of successes in a SRS of size n has approximately the binomial distribution with n equal to the sample size and p equal to the proportion of successes in the population. What are the values of n and p in this example?

$n =$ $p =$

b) You need to use the formula $P(X = k) = \binom{n}{k} p^k (1 - p)^{n-k}$ to evaluate the probability that exactly 2 of the 10 women in the sample have never been married. What are the values of n, k, and p that need to be substituted in the formula and what is the probability requested?

c) Write the probability that 2 or fewer have never been married in terms of X. Your answer should be a sum of binomial probabilities. Be careful rounding off in early stages of the calculations as that may cause substantial errors in your final answer.

d) You can use the formulas that express the mean and standard deviation of X in terms of n and p.

Mean =

Standard deviation =

COMPLETE SOLUTIONS

Exercise 5.19

Although each birth is a boy or girl, we are not counting the number of successes in a fixed number of births. The number of observations (births) is random. The assumption of a fixed number of observations is violated.

Exercise 5.23

You need to evaluate the formula

$$P(X = k) = \binom{n}{k} p^k (1-p)^{n-k},$$

where $n = 20$, $k = 11$, and $p = 20$.

$$P(X = 11) = \binom{20}{11}(0.8)^{11}(0.2)^9 \quad = \frac{20!}{11!9!}(0.8)^{11}(0.2)^9$$

$$= 167960(0.0859)(0.000000512) = 0.0074$$

Exercise 5.29

a) X has a binomial distribution with $n = 1500$ and $p = 0.12$. The mean of X is $np = 1500(0.12) = 180$, and the standard deviation of X is

$$\sqrt{np(1-p)} = \sqrt{1500(0.12)(0.88)} = 12.586$$

b) The normal approximation can be used since both $np = 180$ and $n(1-p) = 1320$ are greater than 10. Using the mean and variance evaluated in part a gives the approximation

$$P(X \le 170) = P\left(\frac{X-180}{12.586} \le \frac{170-180}{12.586}\right) = P(Z \le -0.79) = 0.2148$$

Exercise 5.33

a) X has approximately a binomial distribution with $n = 10$, the sample size, and $p = 0.25$, the proportion of successes (those never having been married) in the population.

b) In this problem, the values that need to be substituted into the formula for evaluating binomial probabilities are $n = 10$, $k = 2$, and $p = 0.25$. This gives

$$P(X = 2) = \binom{10}{2}(0.25)^2(0.75)^8 = \frac{10!}{2!8!}(0.25)^2(0.75)^8 = (45)(0.0625)(0.10012)$$
$$= 0.2816$$

c) You need to evaluate

$$P(X \le 2) = P(X = 0) + P(X = 1) + P(X = 2)$$

$$= \binom{10}{0}(0.25)^0(0.75)^{10} + \binom{10}{1}(0.25)^1(0.75)^9 + \binom{10}{2}(0.25)^2(0.75)^8$$

$$= 0.0563 + 0.1877 + 0.2816 = 0.5256$$

where you need to remember that $0! = 1$ and $(0.25)^0 = 1$ when applying the formulas.

d) The mean of X is $np = 10(0.25) = 2.5$, and the standard deviation of X is

$$\sqrt{np(1-p)} = \sqrt{10(0.25)(0.75)} = 1.37$$

SECTION 5.3

OVERVIEW

Chapter 2 discussed two-way tables and conditional distributions. In this section we learn about **conditional probabilities** and their use in calculating probabilities of complex events. The conditional probability of an event B given that an event A has occurred is denoted $P(B|A)$ and is defined by

$$P(B|A) = \frac{P(A \text{ and } B)}{P(A)}$$

when P(A) > 0. In practice, a conditional probability can often be determined directly from the information given in a problem. Two events A and B are independent if P($B|A$) = P(B).

The **general multiplication rule**, which is valid for events that are not independent, is

$$P(A \text{ and } B) = P(A)P(B|A)$$

This rule can be extended to three events, A, B, and C:

$$P(A \text{ and } B \text{ and } C) = P(A)P(B|A)P(C|A \text{ and } B)$$

GUIDED SOLUTIONS

Exercise 5.43

KEY CONCEPTS - two-way table of counts, conditional probabilities, independence

The table from the problem is reproduced to assist you.

	Bachelor's	Master's	Professional	Doctorates	Total
Female	616	194	30	16	856
Male	529	171	44	26	770
Total	1145	365	74	42	1626

a) You can calculate this probability directly from the table. All degree recipients in the table are equally likely to be selected (that is what it means to select a degree recipient at random), so the fraction of the degree recipients in the table that are women is the desired probability. How many women degree recipients are there? Where do you find this number in the table? What is the total number of degree recipients represented in the table? Use these numbers to compute the desired fraction.

b) This probability can also be calculated directly from the table. Since this is a conditional probability (i.e., this is a probability given that the degree recipient is a professional), we restrict ourselves only to professional degree recipients. The desired probability is then the fraction of these professional degree recipients who are women. Find the appropriate entries in the table to compute this fraction.

c) Recall that two events A and B that both have positive probability are independent if $P(B|A) = P(B)$. Use this rule to determine if the events "choose a woman" and "choose a professional degree recipient" are independent.

COMPLETE SOLUTIONS

Exercise 5.43

a) The number of women degree recipients is found is found as the total for the first row and is (in thousands) 856. The total number of degree recipients in the table is in the lower right corner and is (in thousands) 1626. The desired probability is thus

$$\frac{\text{number of women degree recipients}}{\text{total number of recipients in table}} = \frac{856}{1626} = 0.5624$$

b) The desired probability is

$$\frac{\text{number of professional degree recipients that are women}}{\text{number of professional degree recipients}} = \frac{30}{74} = 0.4054$$

c) In parts a and b we found

$$P(\text{"women is selected"}) = 0.5264$$

$$P(\text{"women is selected"} \mid \text{"professional degree recipient is selected"}) = 0.4054$$

If the events "women is selected" and "professional degree recipient is selected" were independent, these two probabilities would be equal. Since they are not, we conclude that the events are not independent. The proportion of women receiving a professional degree is less than the overall proportion receiving a degree.

SELECTED TEXT REVIEW EXERCISES

GUIDED SOLUTIONS

Exercise 5.55

KEY CONCEPTS - normal approximation to a binomial count, binomial distributions in a SRS

This is an example of using the binomial distribution in the statistical setting of an SRS. When the population is much larger than the sample, a count of successes in an SRS of n has approximately the binomial distribution with n equal to the sample size and p equal to the proportion of successes in the population.

In this example we are counting the number of adults who favor giving parents of school-age children vouchers that can be exchanged for education at any public or private school of their choice, and X is the number in the SRS who favor the vouchers. What are the values of n and p?

   $n =$ $p =$

The probability that more than half the sample favor the vouchers in terms of X is of the form $P(X > c)$ (what is the value of c?). Now you need to use the normal approximation to approximate this probability.

   $np =$ and $n(1 - p) =$

Why can the approximation be used here?

   $P(X > \quad) =$

Exercise 5.63

KEY CONCEPTS - probability rules, conditional probability, interpreting probabilities

You are given information in the problem regarding the probabilities of certain events. The problem first makes sure that you understand the information given, and then has you use it to find other probabilities.

a) Certain of the percentages given are conditional probabilities and others are not. You need to be able to decide this from the language in the problem. Three percentages are given and each is related to probability involving the events A and B. The second two percentages are conditional probabilities. For some of the answers, you may need to use events not A or not B. We have provided the first two answers.

$$0.846 = P(A)$$
$$0.951 = P(B|A)$$
$$0.919 =$$

b) Using the notation in part a, what should be true about $P(B|A)$ and $P(B|\text{not } A)$ if the events A = "a white person is chosen" and B = "an employed person is chosen" were independent? Is this satisfied in this problem? How do you know?

c) You need to use the general multiplication rule here to find

$$P(\text{"employed white person is chosen"}) = P(A \text{ and } B) = P(A)P(B|A)$$

$$=$$

d) You need to use the general multiplication rule here to find

$$P(\text{"employed nonwhite person is chosen"}) =$$

e) $P(\text{"employed person is chosen"}) = P(\text{"employed white person is chosen"})$

$$+ P(\text{"employed nonwhite person is chosen"})$$

Now you can use your answers from parts c and d to get the final answer.

$P(\text{"employed person is chosen"}) =$

COMPLETE SOLUTIONS

Exercise 5.55

We are counting the number of adults who favor giving parents of school-age children vouchers that can be exchanged for education at any public or private school of their choice. X is the number in an SRS who favor the vouchers, so we are counting a success as favoring the voucher. X has approximately a binomial distribution with $n = 500$, the sample size, and $p = 0.45$, the proportion of successes (those favoring the voucher) in the population. If more than half the sample favor the voucher, then $X > 250$, so we need to evaluate the $P(X > 250)$. The normal approximation can be used since both $np = 225$ and $n(1 - p) = 275$ are greater than 10. The mean of X is $np = 500(0.45) = 225$, and the standard deviation of X is

$$\sqrt{np(1-p)} = \sqrt{500(0.45)(0.55)} = 11.124$$

Using the mean and variance just evaluated gives the approximation

$$P(X > 250) = P\left(\frac{X - 250}{11.124} > \frac{250 - 225}{11.124}\right) = P(Z > 2.25) = 1 - 0.9878 = 0.0122$$

Thus the probability that over half the sample favors the vouchers is approximately 0.0122.

Exercise 5.63

a) The second two percentages are conditional probabilities. Of the whites in the labor force we are told that 95.1% are employed. This is the chance of being employed given that you are white, a conditional probability. The last percentage is the same conditional probability, except given that the person is not white (not A).

$$0.846 = P(A)$$
$$0.951 = P(B|A)$$
$$0.919 = P(B|\text{not } A)$$

b) If the events B = "employed person is chosen" and A = "white person is chosen" were independent, then the information about whether or not a person was white wouldn't change the probability that the person was employed. Using the notation in the problem, we should have

$$P(B|A) = P(B|\text{not } A)$$

and these probabilities were shown in part a not to be equal. So the two events are not independent.

c) P("employed white person is chosen")

> $= P$("employed person is chosen"|"white person is chosen")
> $x \, P$("white person is chosen")
> $= (0.951)(0.846) = 0.8045$

using the general multiplication rule. Following the formulas in the book, we are using the rule

$P(A$ and $B) = P(A)P(B|A)$, where A = "a white person is chosen" and B = "employed person is chosen."

d) P("employed nonwhite person is chosen")

> $= P$("employed person is chosen"|"nonwhite is chosen")P("nonwhite is chosen")

> $= (0.919)(1 - 0.846) = 0.1415,$

where we have used the fact that

P("nonwhite is chosen") $= 1 - P$("white is chosen") $= 1 - 0.846 = 0.154$

e) P("employed person is chosen") $= P$("employed white person is chosen")

$+P$("employed nonwhite person is chosen")

$= 0.8045 + 0.1415 = 0.9460,$

where we have used the answers from parts c and d to evaluate the probability.

CHAPTER 6

INTRODUCTION TO INFERENCE

SECTION 6.1

OVERVIEW

A **confidence interval** provides an estimate of an unknown parameter of a population or process, along with an indication of how accurate this estimate is and how confident we are that the interval is correct. Confidence intervals have two parts. One is an interval computed from our data. This interval typically has the form

$$\text{estimate} \pm \text{margin of error}$$

The other part is the **confidence level**, which states the probability that the *method* used to construct the interval will give a correct answer. For example, if you use a 95% confidence interval repeatedly, in the long run 95% of the intervals you construct will contain the correct parameter value. Of course, when you apply the method only once, you do not know if your interval gives a correct value or not. Confidence refers to the probability that the method gives a correct answer in repeated use, not the correctness of any particular interval we compute from data.

Suppose we wish to estimate the unknown mean μ of a normal population with known standard deviation σ based on an SRS of size n. A level C confidence interval for μ is

$$\bar{x} \pm z^* \frac{\sigma}{\sqrt{n}}$$

where z^* is such that the probability is C that a standard normal random variable lies between $-z^*$ and z^* and is obtained from the bottom row in Table C. These z-values are called critical values.

The margin of error $z^* \frac{\sigma}{\sqrt{n}}$ of a confidence interval decreases when any of the following occur:

- The confidence level C decreases.
- The sample size n increases.
- The population standard deviation σ decreases.

The sample size needed to obtain a confidence interval for a normal mean of the form

$$\text{estimate} \pm \text{margin of error}$$

with a specified margin of error m is

$$n = \left(\frac{z^* \sigma}{m} \right)^2$$

where z^* is the critical value for the desired level of confidence. Many times, the n you will find will not be an integer. If it is not, round up to the next larger integer.

The formula for any specific confidence interval is a recipe that is correct under specific conditions. The most important conditions concern the methods used to produce the data. Many methods (including those discussed in this section) assume that our data were collected by random sampling. Other conditions, such as the actual distribution of the population, are also important.

GUIDED SOLUTIONS

Exercise 6.1

KEY CONCEPTS - confidence intervals, interpreting statistical confidence

a) The general form of a confidence interval is

$$\text{estimate} \pm \text{margin of error}$$

Identify the estimate (percentage of the women who said they do not get enough time for themselves) and the margin of error. Then combine these in the general form of a confidence interval as indicated.

b) In formulating your explanation, remember the notions of sampling distributions and sampling variability as discussed in Chapter 4. Write your explanation.

c) In formulating your explanation, consider the meaning of statistical confidence as described in Section 6.1 of the text or the overview for this section of the Study Guide. Write your explanation.

Exercise 6.9

KEY CONCEPTS - confidence intervals for means, the effect of sample size on the margin of error

a) First identify the following quantities. You will want to use the bottom row of Table C to find z^*.

$\bar{x} =$

$\sigma =$

$n =$

z^* (for a 95% confidence interval) =

Now use the formula for a 95% confidence interval to compute the desired interval

$$\bar{x} \pm z^* \frac{\sigma}{\sqrt{n}} =$$

b) Use the same formula as in part a, but now with $n = 250$ rather than 1077.

$$\bar{x} \pm z^* \frac{\sigma}{\sqrt{n}} =$$

c) Once again, use the same formula as in part a, but now with $n = 4000$.

$$\bar{x} \pm z^* \frac{\sigma}{\sqrt{n}} =$$

d) The margins of errors are the quantities after the $\pm$. You computed these in parts a, b, and c. List them here.

Margin of error for $n = 250$:

Margin of error for $n = 1077$:

Margin of error for $n = 4000$:

What pattern do you observe?

Exercise 6.18

KEY CONCEPTS - confidence intervals for means, interpreting statistical confidence

a) Read the statement of the problem carefully, and then complete the following.

The population the authors of the study want to draw conclusions about is:

The population we can be certain the authors can draw conclusions about is:

b) To construct the desired intervals, use the formula

$$\bar{x} \pm z^* \frac{\sigma}{\sqrt{n}}$$

but replace σ with s. What is s here? Write the confidence intervals in the spaces indicated below.

Food stores: $\bar{x} \pm z^* \dfrac{s}{\sqrt{n}} =$

Mass merchandisers: $\bar{x} \pm z^* \dfrac{s}{\sqrt{n}} =$

Pharmacies: $\bar{x} \pm z^* \dfrac{s}{\sqrt{n}} =$

c) In answering this question, consider whether the intervals you computed in part a overlap. Also, bear in mind your answer to part a (to what population do the results apply and to what population do you think the authors want the results to apply?).

Exercise 6.19

KEY CONCEPTS - confidence intervals for means, the effect of changing the confidence level

a) Make your stemplot with split stems in the space provided below. Refer to Chapter 1 if you need to refresh your memory about stemplots.

```
1 |
1 |
2 |
2 |
3 |
3 |
4 |
```

Are there any outliers? Is there extreme skewness?

b) Identify $\sigma, n, \bar{x}$ (you will need to calculate this), and z^* for a 90% confidence interval (use Table C). Then use the formula to calculate the confidence interval.

$$\bar{x} \pm z^* \frac{\sigma}{\sqrt{n}} =$$

c) What is the relation between the level of confidence and the width of the interval? Why is this relation not surprising?

Exercise 6.21

KEY CONCEPTS - sample size required to obtain a confidence interval of specified margin of error

Refer to Exercise 6.19b for the appropriate values of σ and z^*. What margin of error m is desired? Now complete the following to compute the necessary sample size n.

$$n = \left(\frac{z^* \sigma}{m}\right)^2 =$$

Remember to round up to the nearest integer for your final answer.

Exercise 6.23

KEY CONCEPTS - interpreting confidence intervals

a) Review the meaning of statistical confidence and then restate this in plain language.

b) Does the 95% confidence interval contain 50%? What does this imply?

c) Recall the meaning of probability as discussed in Chapter 4. Does probability apply to the fixed unknown proportion of voters that prefer Carter? Explain.

COMPLETE SOLUTIONS

Exercise 6.1

a) The estimate of interest here (percentage of the women who said they do not get enough time for themselves) is 47%. Since the margin of error for a 95% confidence interval is ± 3%, the 95% confidence interval for the percent of all adult women who think they do not get enough time for themselves is 47% ± 3%, or between 44% and 50%.

b) The value 47% is based on a sample of 1025 women selected at random from all women in the United States (excluding Alaska and Hawaii). Another sample of 1025 women might yield a different percent. Repeated random samples of 1025 women will yield a variety of percents. These values will vary around the true percent of women in the United States (excluding Alaska and Hawaii) who feel they do not get enough time for themselves. No particular sample, however, will necessarily give the true value of this percent. Thus we cannot be sure that 47% is the true percent of women in the United States (excluding Alaska and Hawaii) who feel they do not get enough time for themselves. All we can say is that 47% is likely to be "close" to the true percent.

c) Suppose we take all possible random samples of 1025 women. In each sample, suppose we determine the percent of women who think they do not get enough time for themselves. For each of these percents, suppose we add and subtract the margin of error for a 95% confidence interval. Of the resulting intervals, 95% will contain the actual percent of all adult women in the United States (excluding Alaska and Hawaii) who think they do not get enough time for themselves. This is what we mean by "95% confidence." Note that we do not know if any particular interval (such as the 44% to 50% interval in part a) contains the true value of the percent. The confidence level of 95% refers only to the percent of the intervals produced by all samples that will contain the true percent.

Exercise 6.9

a) We are given that the sample mean $\bar{x}$ is 275, the standard deviation σ is 60, and the sample size n is 1077. Thus a 95% confidence interval for the mean score μ in the population of all young women is

$$\bar{x} \pm z^* \frac{\sigma}{\sqrt{n}} = 275 \pm 1.96 \frac{60}{\sqrt{1077}} = 275 \pm 3.58$$

or 271.42 to 278.58.

b) We simply replace $n = 1077$ by $n = 250$ in our calculations and get

$$\bar{x} \pm z^* \frac{\sigma}{\sqrt{n}} = 275 \pm 1.96 \frac{60}{\sqrt{250}} = 275 \pm 7.44$$

or 267.56 to 282.44.

c) We use $n = 4000$ in our calculations and get

$$\bar{x} \pm z^* \frac{\sigma}{\sqrt{n}} = 275 \pm 1.96 \frac{60}{\sqrt{4000}} = 275 \pm 1.86$$

or 273.14 to 276.86.

d) Margin of error for $n = 250$: ± 7.44

   Margin of error for $n = 1077$: ± 3.58

   Margin of error for $n = 4000$: ± 1.86

We see that as the sample size increases, the margin of error decreases.

Exercise 6.18

a) We would assume that the authors of the study wanted to draw conclusions about the population of all adult American consumers. Since the sample was drawn from the Indianapolis phone directory, we can be certain that they can draw conclusions only about the population of all people listed in this phone directory.

b) We use the formula $\bar{x} \pm z^* \frac{\sigma}{\sqrt{n}}$ but with the sample standard deviation, s, in place of σ. Here the sample size is $n = 201$, and since we want 95% confidence intervals, $z^* = 1.96$. The confidence intervals are

Food stores: $18.67 \pm 1.96 \left(\frac{24.95}{\sqrt{201}} \right) = 18.67 \pm 3.45 = (15.22, 22.12)$

Mass merchandisers: $32.38 \pm 1.96 \left(\frac{33.37}{\sqrt{201}} \right) = 32.38 \pm 4.61 = (27.77, 36.99)$

Pharmacies: $48.60 \pm 1.96 \left(\frac{35.62}{\sqrt{201}} \right) = 48.60 \pm 4.92 = (43.68, 53.52)$

c) Neither of the other two 95% confidence intervals overlap with that for pharmacies. In fact, none is even close to overlapping with the interval for pharmacies. This seems reasonably strong evidence that consumers *from the population of all people listed in the Indianapolis phone directory* think pharmacies offer higher performance than other types of stores. Whether this is strong evidence that *all consumers* think pharmacies offer higher performance than other types of stores is less clear. We would need to decide to what extent the population of all people listed in the Indianapolis phone directory are representative of all consumers in the U.S. Without further information, we would be reluctant to extend our conclusions, or extrapolate, to the population of all U.S. consumers.

Exercise 6.19

a) Here is a stemplot of the data. We have used split stems. There are no outliers. There does seem to be some slight right skewness, but it is certainly not extreme.

```
1 | 124
1 | 8
2 | 2233
2 | 6789
3 | 034
3 | 55
4 | 0
```

b) We are told that $\sigma = 8$ and we know that $n = 18$. From the data we calculate $\bar{x} = 25.67$. From Table C we find $z^* = 1.645$ for a 90% confidence interval. Thus our 90% confidence inteval is

$$\bar{x} \pm z^* \frac{\sigma}{\sqrt{n}} = 25.67 \pm 1.645 \frac{8}{\sqrt{18}} = 25.67 \pm 3.10$$

or (22.57, 28.77).

c) Her interval will be wider. It is not surprising that the confidence interval becomes wider as you increase the confidence level. One would expect to be more confident that your interval includes the population mean as you increase the width of the interval.

Exercise 6.21

We want a 90% confidence interval with a margin of error $m = 1$. As we saw in part b of Exercise 6.19, $\sigma = 8$, and from Table C we find $z^* = 1.645$ for a 90% confidence interval. The formula for the proper sample size n is

$$n = \left(\frac{z * \sigma}{m}\right)^2 = \left(\frac{1.645 \times 8}{1}\right)^2 = 173.2$$

which we round up to $n = 174$.

Exercise 6.23

a) Suppose we listed all possible samples we could obtain by the sampling method used to get the sample actually taken. For each, suppose we calculated the percent of the sample that intended to vote for Carter and then attached the appropriate margin of error (the ± 2% for the sample in the article) to each of these percents. Ninety-five percent of the resulting intervals would contain the true percent of people that intended to vote for Carter. Notice that the 95% tells us the percent of samples collected identically to the one actually taken that would contain the true percent of voters favoring Carter. Any particular sample (such as the one actually taken of 51% ± 2%) either does or does not contain the true percent.

b) The 95% confidence interval for the true percent of voters favoring Carter was 51% ± 2% or (49%, 53%). This interval includes values below 50%, which does not rule out the possibility that less than 50% favor Carter. As a result, the polling organization had to say that the election was too close to call (using a 95% confidence level).

c) The true percent of the voters who favor Carter is not known, but it is a fixed value. This fixed value is either larger than 50% or it is not larger than 50%. There is no probability involved. Thus the politician's question does not really make sense. Most likely, the politician is confusing a lack of knowledge, or uncertainty, about the true value with the notion of probability or chance.

Remember, probabilities in polls refer to the method of selecting a sample. In particular, they refer to whether the method used to select the sample will produce results within some given margin of error of the true value. Probabilities do not refer to the true value itself. Since the true value is unknown (that is why we are taking the sample in the first place), there is always a temptation to think of probabilities as referring to our lack of knowledge about this true value. In this text, avoid this temptation.

Note: There are statisticians who define probability in terms of our lack of knowledge or uncertainty about unknown parameters of a population. This is not the view adopted by most statisticians or this text. If you take additional statistics courses, however, you may run across this alternate view. It is sometimes called the Bayesian or subjective view of probability.

SECTION 6.2

OVERVIEW

Tests of significance and confidence intervals are the two most widely used types of formal statistical inference. A test of significance is done to assess the evidence against the **null hypothesis** H_0 in favor of an **alternative hypothesis** H_a. Typically the alternative hypothesis is the effect that the researcher is trying to demonstrate, and the null hypothesis is a statement that the effect is not present. The alternative hypothesis can be either **one-** or **two-sided**.

Tests are usually carried out by first computing a **test statistic**. The test statistic is used to compute a **P-value**, which is the probability of getting a test statistic at least as extreme as the one observed, where the probability is computed when the null hypothesis is true. The *P*-value provides a measure of how incompatible our data are with the null hypothesis, or how unusual it would be to get data like ours if the null hypothesis were true. Since small *P*-values indicate data that are unusual or difficult to explain under the null hypothesis, we typically reject the null hypothesis in these cases. In this case, the alternative hypothesis provides a better explanation for our data.

Significance tests of the null hypothesis H_0: $\mu = \mu_0$ with either a one- or a two-sided alternative are based on the test statistic

$$z = \frac{\bar{x} - \mu_0}{\sigma / \sqrt{n}}$$

The use of this test statistic assumes that we have an SRS from a normal population with known standard deviation σ. When the sample size is large, the assumption of normality is less critical because the sampling distribution of $\bar{x}$ is approximately normal. *P*-values for the test based on z are computed using Table A.

When the *P*-value is below a specified value α, we say the results are **statistically significant at level** α**,** or we reject the null hypothesis at level α. Tests can be carried out at a fixed significance level by obtaining the appropriate critical value z^* from the bottom row in Table C.

GUIDED SOLUTIONS

Exercise 6.25

KEY CONCEPTS - testing hypotheses about means, *P*-value

a) If the null hypothesis H_0: $\mu = 115$ is true, then scores in the population of older students are normally distributed, with mean $\mu = 115$ and standard deviation $\sigma = 30$. What then is the sampling distribution of $\bar{x}$ - the mean of a sample of size $n = 25$? (We studied the sampling distribution of $\bar{x}$ in Chapter 4.)

$N(115, 6)$

Sketch the density curve of this distribution. Be sure to label the horizontal axis properly.

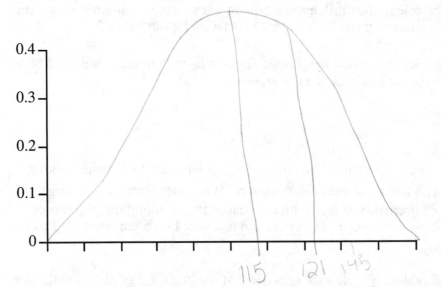

b) Mark the two points on your sketch in part a. Referring to this sketch, explain in simple language why one result is good evidence that the mean score of all older students is greater than 115 and why the other outcome is not. Think about how far out on the density curve the two points are.

c) Shade the appropriate area in your sketch in part a. Refer to Example 6.9 in the text if you need a hint.

Exercise 6.31

KEY CONCEPTS - *P*-values, statistical significance

a) Refer to the graph in your solution to Exercise 6.25. The *P*-value for 118.6 is the shaded area, i.e., the area under the normal curve to the right of 118.6. We learned how to calculate such areas in Chapter 1. First you will need to find the *z* score of 118.6 and then you will need to use Table A to find the area to the right of this *z* score under a standard normal curve. Use the space for your calculations.

Now perform a similar calcualtion to find the *P*-value of 125.7.

b) To answer this question, recall that an observed value is statistically significant at level α if the *P*-value is smaller than α. Use this fact to answer the question.

Exercise 6.37

KEY CONCEPTS - testing hypotheses at a fixed significance level

a) The z-test statistic for testing against a two-sided alternative, as in this problem, is $|z| = \left| \dfrac{\bar{x} - \mu_0}{\sigma / \sqrt{n}} \right|$. Identify μ_0, the standard deviation σ, the sample mean $\bar{x}$, and the sample size n. Then complete the test.

$$|z| = \left| \frac{\bar{x} - \mu_0}{\sigma / \sqrt{n}} \right| =$$

b) To answer you will have to find the appropriate critical value from Table C. Note that we are testing against a two-sided alternative.

c) Follow the procedure in part b, but with significance level 1%.

d) Between what two adjacent critical values in Table C does your z-test statistic lie? What are the corresponding tail probabilities at the top of the table? What do you need to do to the tail probabilities to convert them to critical values for testing against a two-sided alternative?

Exercise 6.39

KEY CONCEPTS - relationship between two-sided tests and confidence intervals

a) You will need to compute $\bar{x}$, then identify σ and the sample size n. Now refer to Section 6.1 for the formula for a 95% confidence interval. Use the space to compute this interval.

b) What must be true of the relationship between a $1-\alpha$ confidence interval for μ and the value μ_0 for a level α two-sided significance test of H_0: $\mu = \mu_0$ to reject the null hypothesis?

Exercise 6.45

KEY CONCEPTS - interpreting P-values

Write your explanation in the space. Refer to the section overview in this Study Guide or the first part of this section in the text if you need a hint.

COMPLETE SOLUTIONS

Exercise 6.25

a) From Chapter 4 we know that the sampling distribution of $\bar{x}$ is normal with mean $\mu = 115$ and standard deviation $\sigma = 30/\sqrt{n} = 30/\sqrt{25} = 30/5 = 6$. A sketch of the density curve of this distribution follows.

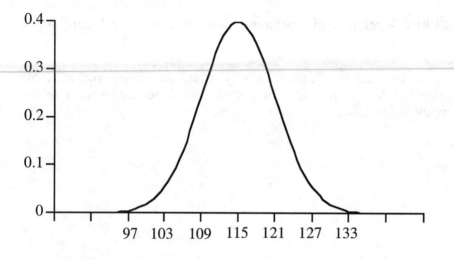

b) The two points are marked on the following curve.

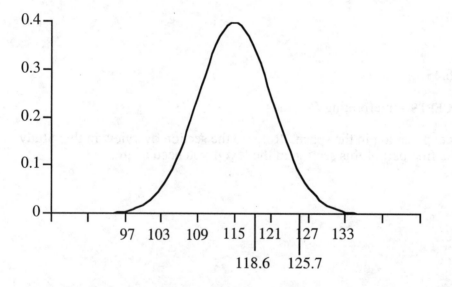

The 125.7 is much farther out on the normal curve than 118.6. In other words, it would be unlikely to observe a mean of 125.7 if the null hypothesis $H_0: \mu = 115$ is true. However, a mean of 118.6 is fairly likely if the null hypothesis is true. A mean as large as 125.7 is more likely to occur if the true mean is larger than 115. Thus, 125.7 is good evidence that the mean score of all older students is greater than 115, while a mean score of 118.6 is not.

c) The *P*-value corresponds to the area to the right of 118.6, since the alternative hypothesis is H_a: $\mu > 115$. This area is displayed in the following figure.

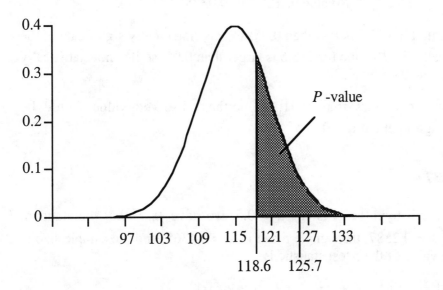

Exercise 6.31

a) Refer to the graph in the complete solution to Exercise 6.25. The *P*-value for 118.6 is the shaded area, i.e., the area under the normal curve to the right of 118.6. We learned how to calculate such areas in Chapter 1. First we compute the *z* score of 118.6. As we saw in Exercise 6.25, the sampling distribution of $\bar{x}$ is normal with mean $\mu = 115$ and standard deviation $\sigma = 6$. We find

$$z \text{ score} = \frac{118.6 - \mu}{\sigma} = \frac{118.6 - 115}{6} = \frac{3.6}{6} = 0.6$$

From Table A we find that the area under the standard normal curve to the right of 0.6 is 1 minus the area to the left of 0.6 = 1 - 0.7527 = 0.2473.

We make a similar calculation for the *P*-value of 125.7:

$$z \text{ score} = \frac{125.7 - \mu}{\sigma} = \frac{125.7 - 115}{6} = \frac{10.7}{6} = 1.78$$

The area to the right of this *z* score under a standard normal curve = 1 minus the area to the left of 1.78 = 1 - 0.9625 = 0.0375.

In summary,

$$P\text{-value of }118.6 = 0.2473$$

$$P\text{-value of }125.7 = 0.0375$$

b) The P-value for 125.7 is less than 0.05, so it is statistically significant at the $\alpha = 0.05$ level. The P-value for 118.6 is larger than 0.05, so it is not statistically significant at the 0.05 level.

Neither P-value is less than 0.01, so neither observed value would be statistically significant at $\alpha = 0.01$.

Exercise 6.37

a) Since the null hypothesis is H_0: $\mu = 0.5$, we have $\mu_0 = 0.5$; the standard deviation is $\sigma = 0.2887$, the sample mean is $\bar{x} = 0.4365$, and the sample size is $n = 100$; the value of the z-test statistic is

$$|z| = \left|\frac{\bar{x} - \mu_0}{\sigma / \sqrt{n}}\right| = \left|\frac{0.4365 - 0.5}{0.2887 / \sqrt{100}}\right| = \left|\frac{-0.0635}{0.02887}\right| = |\text{-}2.20| = 2.20$$

b) Since we are testing against a two-sided alternative, we need to find the upper $\alpha/2 = 0.05/2 = 0.025$ critical value in Table C. We see that this critical value is 1.96. Since our z-test statistic is greater than this critical value, the result is significant at the 0.05 level.

c) Now we look for the upper $\alpha/2 = 0.01/2 = 0.005$ critical value in Table C. We see that this critical value is 2.576. Since our z-test statistic is smaller than this critical value, the result is not significant at the 0.01 level.

d) As we examine the critical values in Table C, we observe that our z-test statistic of 2.20 lies between the critical values 2.054 and 2.326. The corresponding upper tail probabilities at the top of the table are 0.02 and 0.01. Since we are testing against a two-sided alternative, we must multiply the tail probabilities by 2 to get the appropriate critical values (we double the values because we need to include both the upper and lower tail probabilities). Therefore we see that 2.20 lies between the 0.04 and 0.02. We conclude that $0.02 < P\text{-value} < 0.04$.

Exercise 6.39

a) We calculate $\bar{x} = 105.839$. We are given $\sigma = 15$ and n = sample size = 31. From Section 6.1, the formula for our 95% confidence interval is $\bar{x} \pm z^* \dfrac{\sigma}{\sqrt{n}}$, where $z^* = 1.96$ from Table C. We find our 95% confidence interval to be

$$\bar{x} \pm z^* \frac{\sigma}{\sqrt{n}} = 105.839 \pm 1.96 \frac{15}{\sqrt{31}} = 105.839 \pm 5.280$$

or (100.559, 111.119).

b) The hypotheses we wish to test are

$$H_0: \mu = 100 \text{ vs. } H_a: \mu \neq 100$$

where μ is the mean IQ score in the population. Since the value of μ under H_0, namely 100, falls outside the 95% confidence interval we would reject H_0.

Exercise 6.45

In the sample selected by the psychologist, ethnocentrism among church attenders was higher than among nonattenders. Furthermore, the chance of obtaining a difference as large as that observed by the psychologist is less than 0.05 if, in fact, there is no real difference in the population from which the sample was selected. We would take this as strong evidence that ethnocentrism is higher among church attenders than among nonattenders in the population from which the sample was selected.

SECTION 6.3

OVERVIEW

When describing the outcome of a hypothesis test, it is more informative to give the P-value than to just reject or not reject a decision at a particular significance level α. The traditional levels of 0.01, 0.05 and 0.10 are arbitrary and serve as rough guidelines. Different people will insist on different levels of significance depending on the plausibility of the null hypothesis and the consequences of rejecting the null hypothesis. There is no sharp boundary between significant and insignificant, only increasingly strong evidence as the P-value decreases.

When testing hypotheses with a very large sample, the P-value can be very small for effects that may not be of interest. Don't confuse small P-values with large or important effects. Statistical significance is not the same as practical significance. Plot the data to display the effect you are trying to show and also give a confidence interval that says something about the size of the effect.

Statistical inference from data based on a badly designed survey or experiment is often useless. Remember, a statistical test is valid only under certain conditions with data that have been properly produced.

Just because a test is not statistically significant doesn't imply that the null hypothesis is true. Statistical significance may occur when the test is based on a small sample size. Finally, if you run enough tests, you will invariably find statistical significance for one of them. Be careful in interpreting the results when testing many hypotheses on the same data.

GUIDED SOLUTIONS

Exercise 6.55

KEY CONCEPTS - statistical significance versus practical importance

In this problem we see that the P-value associated with the outcome $\bar{x} = 478$ depends on the sample size. The probability of getting a value of $\bar{x}$ as large as 478 if the mean is 475 will become smaller as the sample size gets larger. (Do you remember why? Look at the formula for the z score of $\bar{x}$.) Since this probability is the P-value, we see that a small effect is more likely to be detected for larger sample sizes than for smaller sample sizes. But this doesn't necessarily make the effect interesting or important. A confidence interval tells you something about the size of the effect, not the P-value.

a) Find the P-value by computing the z-test statistic and the probability of exceeding it.

b) This is the same as part a but with a larger sample size. The larger sample size makes the probability of getting a value of $\bar{x}$ as large as 478 smaller than it was in part a. Compute the P-value.

c) Find the P-value in this last case. It will be the smallest. Why?

Exercise 6.57

KEY CONCEPTS - statistical inference is not valid for all sets of data

How was the survey conducted? Would statistical inference be valid for such a survey?

Exercise 6.58

KEY CONCEPTS - multiple analyses

a) What does a *P*-value less than 0.01 mean? Out of 500 subjects, how many would you expect to achieve a score that has such a *P*-value if all 500 are guessing?

b) What would you suggest the researcher now do to test whether any of these four subjects have ESP?

COMPLETE SOLUTIONS

Exercise 6.55

See the guided solutions for a full explanation of the way sample size can change your *P*-values.

a) The *z*-test statistic is

$$z = \frac{\bar{x} - \mu_0}{\sigma / \sqrt{n}} = \frac{478 - 475}{100 / \sqrt{100}} = 0.3$$

and

$$P\text{-value} = P(Z > 0.3) = 1 - 0.6179 = 0.3821$$

since the alternative is one-sided.

b) The test statistic is

$$z = \frac{\bar{x} - \mu_0}{\sigma / \sqrt{n}} = \frac{478 - 475}{100 / \sqrt{1000}} = 0.95$$

and

$$P\text{-value} = P(Z > 0.95) = 1 - 0.8289 = 0.1711$$

c) The test statistic is

$$z = \frac{\bar{x} - \mu_0}{\sigma / \sqrt{n}} = \frac{478 - 475}{100 / \sqrt{10000}} = 3$$

and

$$P\text{-value} = P(Z > 3) = 1 - 0.9987 = 0.0013$$

Exercise 6.57

This was a call-in poll. The data were not collected by a well-designed survey using random sampling. As a result, the confidence interval is probably useless for making inferences about the population of all viewers.

Exercise 6.58

a) A P-value of 0.01 means that the probability a subject would do so well when merely guessing is only 0.01. Among 500 subjects, all of whom are merely guessing, we would therefore expect 1%, or 5, of them to do significantly better than random guessing ($P < 0.01$). Thus in 500 tests it is not unusual to see four results with P-values on the order of 0.01, even if all are guessing and none have ESP.

b) These four subjects only should be retested with a new, well-designed test. If all four again have low P-values (say, below 0.01 or 0.05), we have real evidence that they are not merely guessing. In fact, if any one of the subjects has a very low P-value (say, below 0.01), it would also be reasonably compelling evidence that the individual is not merely guessing. A single P-value on the order of 0.10, however, would not be particularly convincing.

SECTION 6.4

OVERVIEW

From the point of view of making decisions, H_0 and H_a are just two statements of equal status that we must decide between. One chooses a rule for deciding between H_0 and H_a on the basis of the probabilities of the two types of errors we can make. A **Type I error** occurs if H_0 is rejected when it is in fact true. A **Type II error** occurs if H_0 is accepted when in fact H_a is true. There is a clear relation between α-level significance tests and testing from the decision-making point of view. α is the probability of a Type I error.

To compute the Type II error probability of a significance test about a mean of a normal population:

- Write the rule for accepting the null hypothesis in terms of $\bar{x}$.

- Calculate the probability of accepting the null hypothesis when the alternative is true.

The **power** of a significance test is always calculated at a specific alternative hypothesis and is the probability that the test will reject H_0 when that alternative is true. The power of a test against any particular alternative is 1 minus the probability of a Type II error. Power is usually interpreted as the ability of a test to detect an alternative hypothesis or as the sensitivity of a test to an alternative hypothesis. The power of a test can be increased by increasing the sample size when the significance level remains fixed.

GUIDED SOLUTIONS

Exercise 6.63

KEY CONCEPTS - Type I and Type II error probabilities

a) Write the two hypotheses. Remember, we usually take the null hypothesis to be the statement of "no effect."

H_0:

H_a:

Describe the two types of errors as "false positive" and "false negative" test results.

b) Which error probability would you choose to make smaller (at the expense of making the other error probability larger) and why?

Exercise 6.65

KEY CONCEPTS - Type I and Type II error probabilities

a) If $\mu = 0$, what is the sampling distributionn of $\bar{x}$? Now use this to compute the probability the test rejects, i.e., the probability $\bar{x} > 0$.

b) If $\mu = 0.3$, what is the sampling distributionn of $\bar{x}$? Now use this to compute the probability the test accepts H_0, i.e., the probability $\bar{x} \leq 0$.

c) If $\mu = 1$, what is the sampling distributionn of $\bar{x}$? Now use this to compute the probability the test accepts H_0, i.e., the probability $\bar{x} \pm 0$.

Exercise 6.67

KEY CONCEPTS - power

Begin by rewriting the rule for rejecting H_0 in terms of $\bar{x}$. We help you by getting you started. The rule is to reject H_0 if $z \le -1.645$, or

$$z = \frac{\bar{x} - 300}{3/\sqrt{6}} = \frac{\bar{x} - 300}{1.22} \le -1.645.$$

What does this inequality imply about the values of $\bar{x}$?

a) If $\mu = 299$, what is the sampling distribution of $\bar{x}$? Use this to compute the probability that the test rejects, i.e., the probability that $\bar{x}$ takes on values (which you computed above) that lead to rejecting H_0 when the particular alternative $\mu = 299$ is true.

b) If $\mu = 295$, what is the sampling distributionn of $\bar{x}$? Use this to compute the probability that the test rejects, i.e., the probability that $\bar{x}$ takes on values (which you computed above) that lead to rejecting H_0 when the particular alternative $\mu = 295$ is true.

c) What do you notice in parts a and b about the change in power as the true value of μ changes from 299 to 295?

Exercise 6.71

KEY CONCEPTS - power and the relationship with the Type II error probability

What is the relationship between the probability of a Type I error and the level of significance?

What is the relationship between the power of a test at a particular alternative and the Type II error at this alternative? Now use the value of the power that you calculated in Exercise 6.67 to compute the Type II error probability.

COMPLETE SOLUTIONS

Exercise 6.63

a) The two hypotheses are

H_0: the patient has no medical problem
H_a: the patient has a medical problem

One possible error is to decide

H_a: the patient has a medical problem

when, in fact, the patient does not really have a medical problem. This is a Type I error and in this setting could be called a false positive. The other type of error is to decide

$$H_0: \text{the patient has no medical problem}$$

when, in fact, the patient does have a problem. This is a Type II error and in this setting could be called a false negative.

b) Most likely we would choose to decrease the error probability for a Type II error, or the false negative probability. Failure to detect a problem (particularly a major problem) when one is present could result in serious consequences (such as death). While a false positive can also have serious consequences (painful or expensive treatment that is not necessary), it is not likely to lead to the kinds of consequences that a false negative could produce. For example, consider the consequences of failure to detect a heart attack, the presence of AIDS, or the presence of cancer. Note, there are cases where some might argue that a false positive would be a more serious error than a false negative. For example, a false positive in a test for Down's Syndrome or a birth defect in an unborn baby might lead parents to consider an abortion. Some would consider this a much more serious error than to give birth to a child with a birth defect.

Exercise 6.65

a) If $\mu = 0$, the sampling distribution of $\bar{x}$ is normal with mean $\mu = 0$ and standard deviation $\dfrac{\sigma}{\sqrt{n}} = \dfrac{1}{\sqrt{9}} = 0.33$. Thus the probability of a Type I error is the probability that $\bar{x} > 0$ when the null hypothesis is true. Computing the z score for $\bar{x}$, we get

$$P(\bar{x} > 0) = P\left(\frac{\bar{x} - 0}{0.33} > \frac{0 - 0}{0.33}\right) = P(Z > 0) = 0.5$$

b) If $\mu = 0.3$, the sampling distribution of $\bar{x}$ is normal with mean $\mu = 0.3$ and standard deviation $\dfrac{\sigma}{\sqrt{n}} = \dfrac{1}{\sqrt{9}} = 0.33$. We accept H_0 if $\bar{x} \leq 0$. Thus the probability of a Type II error when $\mu = 0.3$ is

$$P(\bar{x} \leq 0) = P\left(\frac{\bar{x} - 0.3}{0.33} \leq \frac{0 - 0.3}{0.33}\right) = P(Z \leq -0.91) = .1814$$

c) If $\mu = 1$, the sampling distribution of $\bar{x}$ is normal with mean $\mu = 1$ and standard deviation $\dfrac{\sigma}{\sqrt{n}} = \dfrac{1}{\sqrt{9}} = 0.33$. We accept H_0 if $\bar{x} \leq 0$. Thus the probability of a Type II error when $\mu = 1$ is

$$P(\bar{x} \le 0) = P\left(\frac{\bar{x}-1}{0.33} \le \frac{0-1}{0.33}\right) = P(Z \ -3.0) = .0013$$

Exercise 6.67

We begin by rewriting the rule for rejecting H_0 in terms of $\bar{x}$. The rule is to reject H_0 if $z \le -1.645$, or

$$\frac{\bar{x} - 300}{3/\sqrt{6}} = \frac{\bar{x} - 300}{1.22} \le -1.645.$$

Rewriting this in terms of $\bar{x}$ gives the following rule for rejection. Reject H_0 if

$$\bar{x} \le (1.22)(-1.645) + 300 = 297.99$$

a) If $\mu = 299$, the sampling distribution of $\bar{x}$ is normal with mean 299 and standard deviation $\frac{3}{\sqrt{6}} = 1.22$. The power is the probability of rejecting H_0 when the particular alternative $\mu = 299$ is true. This probability is

$$\text{Power} = P(\bar{x} \le 297.99) = P\left(\frac{\bar{x} - 299}{1.22} \le \frac{297.99 - 299}{1.22}\right)$$

$$= P(Z \le -0.83) = .2033$$

b) If $\mu = 295$, the sampling distribution of $\bar{x}$ is normal with mean 295 and standard deviation $\frac{3}{\sqrt{6}} = 1.22$. The power is the probability of rejecting H_0 when the particular alternative $\mu = 295$ is true. This probability is

$$\text{Power} = P(\bar{x} \le 297.99) = P\left(\frac{\bar{x} - 295}{1.22} \le \frac{297.99 - 295}{1.22}\right)$$

$$= P(Z \le 2.45) = .9929$$

c) The power will be greater than in part b. We notice in parts a and b that as μ decreased, the power increased. Thus we would suspect that the power will be greater when μ decreases further to 290. A more precise argument is the following. If $\mu = 290$, $\bar{x}$ is likely to be close to 290 (within a standard deviation or two of 290). 297.99 is several multiples of the standard deviation of 1.22 above 290. Thus it is almost certain that $\bar{x}$ will be less than 297.99 and more likely to be less than 297.99 than if μ was 295 (which is closer to 297.99 than 290 is). This means the power will be very close to 1 and greater than in part b.

Exercise 6.71

Recall that the test in Exercise 6.67 used a significance level of 5% to test the hypotheses

$$H_0: \mu = 300$$
$$H_a: \mu < 300$$

The probability of a Type I error is the same as the significance level and hence is 0.05.

The probability of a Type II error at the alternative $\mu = 295$ is the probability of accepting the null hypothesis H_0: $\mu = 300$. One minus this probability is the probability of (correctly) rejecting the null hypothesis H_0: $\mu = 300$ when the alternative $\mu = 295$ is true. This last probability is the power at the alternative $\mu = 295$. We found this to be .9929 in part b of exercise 6.67. One minus this value is thus the Type II error. Hence the Type II error is 1 - .9929 = .0071.

SELECTED TEXT REVIEW EXERCISES

GUIDED SOLUTIONS

Exercise 6.81

KEY CONCEPTS - interpreting *P*-values

Write your explanation in the space provided. Refer to the section overview in the Study Guide or the first part of this section in the text if you need to refresh your memory as to the proper interpretation of a *P*-value.

COMPLETE SOLUTIONS

Exercise 6.81

This is not a correct explanation. Remember, the null hypothesis is either true or false. There is no probability involved. Probability enters when we talk about the possible values we will obtain from a sample that will be used to test the null hypothesis. Probability refers to the chance our sample will give us reliable or misleading information. A correct explanation of the P-value of .03 is that if the null hypothesis is true, there is only a probability of 3% that we would observe a sample result as inconsistent with the null hypothesis as actually observed, by chance. In other words, P-values tell us the probability that our sample results are due to chance (assuming that the null hypothesis is true) as opposed to being the result of a real effect.

CHAPTER 7

INFERENCE FOR DISTRIBUTIONS

SECTION 7.1

OVERVIEW

Confidence intervals and significance tests for the mean μ of a normal population are based on the sample mean $\bar{x}$ of an SRS. When the sample size n is large, the central limit theorem suggests that these procedures are approximately correct for other population distributions. In Chapter 6, we considered the (unrealistic) situation in which we knew the population standard deviation σ. In this section, we consider the more realistic case where σ is not known and we must estimate σ from our SRS by the sample standard deviation s. In Chapter 6 we used the **one-sample z statistic**

$$z = \frac{\bar{x} - \mu}{\sigma/\sqrt{n}}$$

which has the $N(0,1)$ distribution. Replacing σ by s, we now use the **one-sample t statistic**

$$t = \frac{\bar{x} - \mu}{s/\sqrt{n}}$$

which has the *t* **distribution** with *n* - 1 **degrees of freedom**.

For every positive value of *k* there is a *t* distribution with *k* degrees of freedom, denoted $t(k)$. All are symmetric, bell-shaped distributions, similar in shape to normal distributions but with greater spread. As *k* increases, $t(k)$ approaches the $N(0,1)$ distribution.

A level *C* **confidence interval for the mean** μ of a normal population when σ is unknown is

$$\bar{x} \pm t^* \frac{s}{\sqrt{n}}$$

where t^* is the upper $(1 - C)/2$ critical value of the $t(n - 1)$ distribution whose value can be found in Table C of the text or from statistical software. The one-sample *t* confidence interval has the form

$$\text{estimate} \pm t^* \text{SE}_{\text{estimate}}$$

where SE stands for **standard error**.

Significance tests of $H_0: \mu = \mu_0$ are based on the one-sample *t* statistic. *P*-values or fixed significance levels are computed from the $t(n - 1)$ distribution using Table C or, more commonly in practice, using statistical software.

One application of these one-sample *t* procedures is to the analysis of data from **matched pairs** studies. We compute the differences between the two values of a matched pair (often before and after measurements on the same unit) to produce a single sample value. The sample mean and standard deviation of these differences are computed. Depending on whether we are interested in a confidence interval or a test of significance concerning the difference in the population means of matched pairs, we use either the one-sample confidence interval or the one-sample significance test based on the *t* statistic.

For larger sample sizes, the *t* procedures are fairly **robust** against nonnormal populations. As a rule of thumb, *t* procedures are useful for nonnormal data when $n \geq 15$ unless the data show outliers or strong skewness, and for samples of size $n \geq 40$ *t* procedures can be used for even clearly skewed distributions. For smaller samples, it is a good idea to examine stemplots or histograms before you use the *t* procedures to check for outliers or skewness.

GUIDED SOLUTIONS

Exercise 7.5

KEY CONCEPTS - critical values of the $t(k)$ distribution, P-values for t procedures

a) For confidence intervals or significance tests, the critical values for the one-sample t procedures use the $t(n - 1)$ distribution. What are the degrees of freedom for the t statistic given in this problem?

b) Between what two critical values t^* does $t = 1.82$ fall? What are the right-tail probabilities corresponding to these entries? In which row of Table C do you need to look to answer these questions?

c) The P-value is $P(T \geq 1.82)$, since the alternative is $H_a: \mu > 0$. Between what two values does the P-value lie?

d) Is the value $t = 1.82$ significant at the 5% level? Why or why not?

What about at the 1% level?

Exercise 7.9

KEY CONCEPTS - matched pairs experiments, one-sample t tests

a) This is a matched pairs experiment. The matched pair of observations are the right- and left-hand times on each subject. To avoid confounding with time of day, we would probably want subjects to use both knobs in the same session. We would also want to randomize which knob the subject uses first. How might you do this randomization? What about the order in which the subjects are tested?

b) The project hopes to show that right-handed people find right-hand threads easier to use than left-hand threads. In terms of the mean μ for the population of differences

(left thread time) - (right thread time)

what do we wish to show? This would be the alternative. What are H_0 and H_a (in terms of μ)?

H_0:
H_a:

c) For data from a matched pairs study, we compute the differences between the two values of a matched pair to produce a single sample value. These differences are given below for our data.

Right thread	Left thread	Difference = Left - Right
113	137	24
105	105	0
130	133	3
101	108	7
138	115	-23
118	170	52
87	103	16
116	145	29
75	78	3
96	107	11
122	84	-38
103	148	45
116	147	31
107	87	-20
118	166	48
103	146	43
111	123	12
104	135	31
111	112	1
89	93	4
78	76	-2
100	116	16
89	78	-11
85	101	16
88	123	35

Compute the sample mean and standard deviation of these differences.

$\bar{x} = $ $s = $

We now use the one-sample significance test based on the t statistic. What value of μ_0 should be used?

$$t = \frac{\bar{x} - \mu_0}{s / \sqrt{n}} =$$

From the value of the t statistic and Table C (or using statistical software), the P-value can be computed. If using Table C, between what two values does the P-value lie?

$\leq P\text{-value} \leq$

What conclusion do you draw?

Note: This problem is most easily done directly using statistical software. The software will compute the differences, the t statistic, and the P-value for you. Consult your users' manual to see how to do one-sample t tests.

Exercise 7.13

a) Recall some of the graphical methods of Chapter 1 for describing data. Examine the boxplot. Complete the histogram using the class intervals provided.

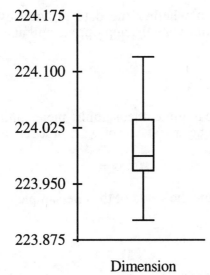

Dimension

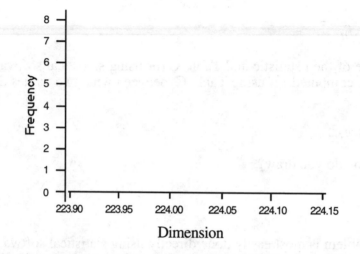

Using the histogram and/or the boxplot, is there evidence for outliers or strong skewness? Given the general guidelines on the robustness of the t procedures, do you think it is valid to use these procedures for this data?

b) We are interested in whether the data provide evidence that the mean dimension is not 224 mm. State the appropriate null and alternative hypotheses.

H_0:
H_a:

From the data, first calculate the sample mean and the sample standard deviation, preferably using statistical software or a calculator.

$\bar{x} =$ $\qquad$ $s =$

Now use them to compute the value of the one-sample t statistic.

$$t = \frac{\bar{x} - \mu_0}{s / \sqrt{n}} =$$

Compute the *P*-value using Table C or software. How many degrees of freedom are there? The degrees of freedom tell you which row of Table C you need to refer to for critical values. Remember, if the alternative is two-sided, then the probability found in the table needs to be doubled.

Degrees of freedom:

P-value:

Conclusion:

Exercise 7.17

KEY CONCEPTS - confidence intervals based on the one-sample *t* statistic, assumptions underlying *t* procedures

a) To compute a level *C* confidence interval we use the formula $\bar{x} \pm t^* \frac{s}{\sqrt{n}}$, where t^* is the upper (1 - *C*)/2 critical value of the *t*(*n* - 1) distribution, which can be found in Table C. Fill in the values. Don't forget to subtract 1 from the sample size when finding the appropriate degrees of freedom for the *t* confidence interval.

$C =$
$n =$
$t^* =$

The values of $\bar{x}$ and *s* are given in the problem.

$\bar{x} =$ $\qquad\qquad\qquad\qquad$ $s =$

Substitute all these values into the formula to complete the computation of the 95% confidence interval.

$$\bar{x} \pm t^* \frac{s}{\sqrt{n}} =$$

b) What are the assumptions required for the *t* confidence interval? Which assumptions are satisfied and which may not be? How were the subjects in the study obtained? How were the subjects in the placebo group obtained?

Exercise 7.25

KEY CONCEPTS - appropriateness of statistical procedures

Statistical procedures use information in a sample to make inferences about parameters of a population. Ask yourself, What is the population and the parameter of interest in this exercise? What is the sample?

COMPLETE SOLUTIONS

Exercise 7.5

a) The degrees of freedom for the one-sample t procedures are $n - 1 = 15 - 1 = 14$.

b) You need to refer to the row corresponding to 14 degrees of freedom. The value of $t = 1.82$ falls between the two t^* critical values 1.761 and 2.145, corresponding to upper tail probabilities of 0.05 and 0.025, respectively.

df = 14

p	.05	.025
t^*	1.761	2.145

c) The $P(T \geq 1.82)$ must lie between 0.025 and 0.05. (If the alternative were two-sided, these values would need to be doubled).

d) The value $t = 1.82$ is signficant at the 5% level since the P-value is below 0.05, but not at the 1% level since the P-value exceeds 0.01.

Exercise 7.9

a) The randomization might be carried out by simply flipping a fair coin. If the coin comes up heads, use the right-threaded knob first. If the coin comes up tails, use the left-threaded knob first. Alternatively, in order to balance out the number of times each type is used first, one might choose an SRS of 12 of the 25 subjects. These 12 use the right-threaded knob first. Everyone else uses the left-threaded knob first.

A second place one might use randomization is in the order in which subjects are tested. Use a table of random digits to determine the order. Label subjects 01 to 25. The first label that appears in the list of random digits (read in groups of two digits) is the first subject measured. The second label that appears, the next subject measured, and so on. This randomization is probably less important than the one described in the previous paragraph. It would be important if the order or time at which a subject was tested has an effect on the measured response. For example, if the study began early in the morning, the first subject might be sluggish if still sleepy. Sluggishness might lead to longer times and perhaps a larger difference in times. Subjects tested later in the day might be more alert.

b) In terms of μ, the mean of the population of differences, (left thread time) - (right thread time), we wish to test if the times for the left-threaded knobs are longer than for the right-threaded knobs, i.e.,

$$H_0: \mu = 0 \text{ vs. } H_a: \mu > 0$$

c) For the 25 differences we compute

$$\bar{x} = 13.32 \qquad\qquad s = 22.94$$

We then use the one-sample significance test based on the t statistic.

$$t = \frac{\bar{x} - \mu_0}{s / \sqrt{n}} = \frac{13.32 - 0}{22.94 / \sqrt{25}} = 2.903$$

From the value of the t statistic and Table C the P-value is between 0.0025 and 0.005.

df = 24

p	.005	.0025
$t*$	2.797	3.091

Using statistical software, the P-value is computed as P-value = 0.0039.

We conclude that there is strong evidence that the time for left-hand threads is greater than the time for right-hand threads on average.

Exercise 7.13

a) The boxplot was provided in the guided solutions, and the completed histogram is given here (one could also make a stemplot of the data).

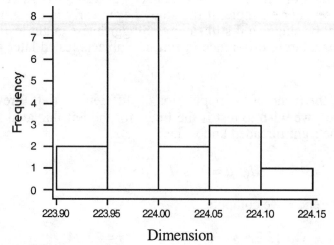

Dimension

Using either the histogram or the boxplot, we can see that there are no outliers in the data. The data appear a bit skewed to the right, but not strongly enough to threaten the validity of the t procedure given that the sample size is 16 (in the section on the robustness of t procedures, t procedures are safe for samples of size $n \geq 15$ unless there are outliers and/or strong skewness).

b) Since we are interested in whether the data provide evidence that the mean dimension is not 224 mm (no direction of the difference is specified), we wish to test the hypotheses

H_0: $\mu = 224$ mm

H_a: $\mu \neq 224$ mm

From the data we calculate the basic statistics to be

$\bar{x} = 224.0019$, $s = 0.618$, and the standard error as $\dfrac{s}{\sqrt{n}} = \dfrac{0.0618}{\sqrt{16}} = 0.01545$.

Substituting these in the formula for t yields

$$t = \frac{\bar{x} - \mu_0}{s / \sqrt{n}} = \frac{224.0019 - 224}{0.01545} = 0.123$$

The P-value for $t = 0.123$ is the twice the area to the right of 0.123 under the t distribution curve with $n - 1 = 15$ degrees of freedom. Using Table C, we search the df = 15 row for entries that bracket 0.123. Since 0.123 lies to the left of the smallest entry in the table corresponding to a probability of 0.25, the P-value is therefore greater than .25 x 2 = .50 for this two-sided test. The data do

$$df = 15$$

p	.25
t^*	0.691

not provide strong evidence that the mean differs from 224 mm (computer software gives a P-value of 0.9038).

Exercise 7.17

a) A 95% confidence interval for the mean systolic blood pressure in the population from which the subjects were recruited can be calculated from the data on the 27 members of the placebo group, since these are randomly selected from the 54 subjects. We use the formula for a t interval, namely $\bar{x} \pm t^* \dfrac{s}{\sqrt{n}}$. In this problem, $\bar{x} = 114.9$, $s = 9.3$, $n = 27$, hence t^* is the upper $(1 - 0.95)/2 = 0.025$ critical value for the $t(26)$ distribution. From Table C we see that $t^* = 2.056$. Thus the 95% confidence interval is

$$114.9 \pm 2.056 \frac{9.3}{\sqrt{27}} = 114.9 \pm 3.68 = (111.22, 118.58)$$

b) For the procedure used in part a, the population from which the subjects were drawn should be such that the distribution of the seated systolic blood pressure in the population is normal. The 27 subjects used for the confidence interval in part a should be a random sample from this population. Unfortunately, we do not know if the latter is the case. While 27 subjects were selected at random from the total of 54 subjects in the study, we do not know if the 54 subjects were a random sample from this population.

With a sample of 27 subjects, it is not crucial that the population be normal, as long as the distribution is not strongly skewed and the data contain no outliers. It is important that the 27 subjects can be considered a random sample from the population. If not, we cannot appeal to the central limit theorem to ensure that the t procedure is at least approximately correct even if the data are not normal.

(Note: It turns out that since the subjects were divided at random into treatment and control groups, there do exist procedures for comparing the treatment and placebo groups. These are not based on the t distribution but are valid as long as treatment groups are determined by randomization. However, the conclusions drawn from these procedures apply only to the subjects in the study. To generalize the conclusions to a larger population, we must know that the subjects are a random sample from the larger population.)

Exercise 7.25

We do not have a sample of U.S. presidents. We have the ages of the entire *population* of U.S. presidents. The population average age is known exactly by simply averaging the numbers in Table 1.7. A confidence interval is not necessary, since there is no uncertainty concerning the value of the population mean.

SECTION 7.2

OVERVIEW

One of the the most commonly used significance tests is the **comparison of two population means,** μ_1 and μ_2. In this setting we have two distinct, independent SRSs from two populations or two treatments in a randomized comparative experiment. The procedures are based on the difference $\bar{x}_1 - \bar{x}_2$. When the populations are not normal, the results obtained using the methods of this section are approximately correct due to the central limit theorem.

Tests and confidence intervals for the difference in the population means, $\mu_1 - \mu_2$, are based on the **two-sample t statistic**. Despite the name, this test statistic does *not* have an exact t distribution. However there are good approximations to its distribution that allow us to carry out valid significance tests. Conservative procedures use the $t(k)$ distribution as an approximation, where the degrees of freedom k is taken to be the smaller of $n_1 - 1$ and $n_2 - 1$. More accurate procedures use the data to estimate the degrees of freedom k. This is the procedure that is followed by most statistical software.

To carry out a significance test for $H_0: \mu_1 = \mu_2$, use the two-sample t statistic

$$t = \frac{(\bar{x}_1 - \bar{x}_2)}{\sqrt{\dfrac{s_1^2}{n_1} + \dfrac{s_2^2}{n_2}}}$$

The P-value is found by using the approximate distribution $t(k)$, where k is estimated from the data when using statistical software or can be taken to be the smaller of $n_1 - 1$ and $n_2 - 1$ for a conservative procedure.

An approximate level C **confidence interval** for $\mu_1 - \mu_2$ is given by

$$(\bar{x}_1 - \bar{x}_2) \pm t^* \sqrt{\frac{s_1^2}{n_1} + \frac{s_2^2}{n_2}}$$

where t^* is the upper $(1 - C)/2$ critical value for $t(k)$, where k is estimated from the data when using statistical software, or can be taken to be the smaller of $n_1 - 1$ and $n_2 - 1$ for a conservative procedure. The procedures are most robust to failures in the assumptions when the sample sizes are equal.

The **pooled two-sample t procedures** are used when we can safely assume that the two populations have equal variances. The modifications in the procedure are the use of the pooled estimator of the common unknown variance and critical values obtained from the $t\,(n_1 + n_2 - 2)$ distribution.

GUIDED SOLUTIONS

Exercise 7.29

KEY CONCEPTS - single sample, matched pairs, or two samples

a) Are there one or two samples involved? Was matching done?

b) Are there one or two samples involved? Was matching done?

Exercise 7.37

KEY CONCEPTS - confidence intervals and tests using the two-sample t with degrees of freedom estimated from the data

a) We are interested in whether the data provide evidence that poisoned rats differ from unpoisoned rats. State the appropriate null and alternative hypotheses. Let μ_{DDT} and $\mu_{Control}$ correspond to the population mean of the measured variable for the two groups.

   H_0:
   H_a:

b) In the SAS output, we are interested in the output for the general two-sample procedure for "unequal" variances. The t is computed as in the text, but the degrees of freedom are estimated from the data using the formula in the text box "Approximate Distribution of the Two-sample t Statistic." The degrees of freedom is generally not an integer, but this is not a problem when the computer is doing the arithmetic.

What is the value of the two-sample t statistic and its P-value? What do you conclude?

c) The formula for the 90% confidence interval is $\bar{x}_1 - \bar{x}_2 \pm t^* \sqrt{\dfrac{s_1^2}{n_1} + \dfrac{s_2^2}{n_2}}$, where

t^* is the upper $(1 - C)/2 = 0.05$ critical value for the t distribution with degrees of freedom equal to 5.9. Since the degrees of freedom is not a whole number, we use the next smaller degrees of freedom, which is 5. The means and standard deviations are given in the output. Don't forget to square the standard deviations in the formula for the standard error. Complete the calculations in steps.

$t^* =$

$$\sqrt{\frac{s_1^2}{n_1} + \frac{s_2^2}{n_2}} =$$

$$\bar{x}_1 - \bar{x}_2 \pm t^* \sqrt{\frac{s_1^2}{n_1} + \frac{s_2^2}{n_2}} =$$

Exercise 7.49

KEY CONCEPTS - tests and confidence intervals using the two-sample t, relationship between two-sided tests and confidence intervals

Following are the summary statistics for the comparison of children and adult VOT scores.

Group	n	$\bar{x}$	s
Children	10	-3.67	33.89
Adults	20	-23.17	50.74

a) Were the researchers interested in a difference in either direction? State the appropriate null and alternative hypotheses.

$$H_0:$$

$$H_a:$$

Now find the numerical value of the two-sample t statistic using the following formula. Don't forget to square both standard deviations when evaluating the formula for the standard error in the denominator of the t.

$$t = \frac{\bar{x}_1 - \bar{x}_2}{\sqrt{\dfrac{s_1^2}{n_1} + \dfrac{s_2^2}{n_2}}} =$$

Compute the P-value using a t distribution with the smaller of $n_1 - 1 = 9$ and $n_2 - 1 = 19$ degrees of freedom, and then compare the computed value of t to the critical values given in Table C. Remember, if the test is two-sided you will need to double the tail probabilities. State your conclusions.

b) The formula for the 95% confidence interval is $\bar{x}_1 - \bar{x}_2 \pm t^* \sqrt{\dfrac{s_1^2}{n_1} + \dfrac{s_2^2}{n_2}}$, where t^* is the upper $(1 - C)/2 = 0.025$ critical value for the t distribution with degrees of freedom equal to the smaller of $n_1 - 1 = 9$ and $n_2 - 1 = 19$. First find t^* from Table C, then complete the computations for the confidence interval.

Remember the relationship between two-sided tests and confidence intervals. We reject the null hypothesis $H_0: \mu_1 = \mu_2$ at significance level α when the $1 - \alpha$ confidence interval for $\mu_1 - \mu_2$ doesn't contain the value 0. How did you know from your result in part a that the interval would contain 0 (no difference)?

COMPLETE SOLUTIONS

Exercise 7.29

a) This example involves a single sample. We have a sample of 20 measurements and we want to see whether the mean for this method agrees with the known concentration.

b) This example involves two samples, the set of measurements on each method. Note that we are not told of any matching.

Exercise 7.37

a) Since we are interested in whether poisoned rats differ from unpoisoned rats, we are interested in differences in either direction. The appropriate hypotheses are H_0: $\mu_{\text{DDT}} = \mu_{\text{Control}}$ and H_a: $\mu_{\text{DDT}} \neq \mu_{\text{Control}}$.

b) Using the line for unequal variances, we have $t = 2.9912$, df $= 5.9$ and the P-value $= 0.0247$. We would reject the null hypothesis at a significance level of 0.05, since the P-value falls below 0.05, but not at the 0.01 significance level. We conclude that there is good evidence of a difference in the mean of the measured variable (height of the second spike as a percent of the first) between poisoned and unpoisoned rats.

c) The 0.05 critical value of the $t(5)$ distribution is 2.015 from Table C. The standard error is

$$\sqrt{\frac{s_1^2}{n_1} + \frac{s_2^2}{n_2}} = \sqrt{\frac{6.34^2}{6} + \frac{1.95^2}{6}} = \sqrt{7.33} = 2.708$$

and the 90% confidence interval is

$$\bar{x}_1 - \bar{x}_2 \pm t^* \sqrt{\frac{s_1^2}{n_1} + \frac{s_2^2}{n_2}} = 17.60 - 9.50 \pm 2.015(2.708) = 8.10 \pm 5.46$$

Since the test rejected H_0 at the 0.10 level of significance, we know that the 90% confidence interval would not contain 0.

Exercise 7.49

a) The hypotheses are H_0: $\mu_1 = \mu_2$ and H_a: $\mu_1 \neq \mu_2$, since the researchers were interested in a difference in either direction (do VOT scores distinguish adults from children?). The standard error for the difference $\bar{x}_1 - \bar{x}_2$ between the mean VOT for children and adults is

$$\sqrt{\frac{s_1^2}{n_1} + \frac{s_2^2}{n_2}} = \sqrt{\frac{33.89^2}{10} + \frac{50.74^2}{20}} = 15.607$$

Substituting this value into the denominator of the formula for t yields

$$t = \frac{\bar{x}_1 - \bar{x}_2}{\sqrt{\frac{s_1^2}{n_1} + \frac{s_2^2}{n_2}}} = \frac{-3.67 - (-23.17)}{15.607} = 1.25$$

The smaller of $n_1 - 1 = 9$ and $n_2 - 1 = 19$ is 9, so we can refer the computed value of t to a $t(9)$ in Table C. The value 1.25 is between the critical values corresponding to upper tail probablities of 0.10 and 0.15. This gives a P-value between 0.20 and 0.30, since we need to double upper tail probabilities from the table because the test is two-sided (the exact P-value for a t with 9 degrees of freedom using statistical software is 0.2428). The data give no evidence of difference in mean VOT between children and adults.

b) The 95% confidence interval is

$$\bar{x}_1 - \bar{x}_2 \pm t^* \sqrt{\frac{s_1^2}{n_1} + \frac{s_2^2}{n_2}}$$

where t^* is the upper $(1 - C)/2 = 0.025$ critical value for the t distribution with degrees of freedom equal to the smaller of $n_1 - 1 = 9$ and $n_2 - 1 = 19$. Using 9 degrees of freedom, we find the value of $t^* = 2.262$, and the confidence interval is

$$-3.67 - (-23.17) \pm 2.262(15.607) = (-15.8, 54.8)$$

Since the interval includes the value 0, we fail to reject H_0 at the 5% level of significance. We know this from part a since the P-value exceeded 5%.

SECTION 7.3

OVERVIEW

There are formal inference procedures for comparing the standard deviations of two normal populations as well as the two means. The validity of the procedures is seriously affected when the distributions are nonnormal, and they are not recommended for regular use. The procedures are based on the **F statistic,** which is the ratio of the two sample variances:

$$F = \frac{s_1^2}{s_2^2}$$

If the data consist of independent simple random samples of sizes n_1 and n_2 from two normal populations, then the F statistic has the F distribution, $F(n_1 - 1, n_2 - 1)$, if the two population standard deviations σ_1 and σ_2 are equal.

Critical values of the F distribution are provided in Table D. Because of the skewness of the F distribution, when carrying out the two-sided test we take the ratio of the larger to the smaller standard deviation, which eliminates the need for lower critical values.

GUIDED SOLUTIONS

Exercise 7.57

KEY CONCEPTS - F test for equality of the standard deviations of two normal populations

The data are from Exercise 7.49 and are measurements on VOT (voice onset time) for a sample of children and adults. The summary statistics for the comparison of children and adult VOT scores follow.

Group	n	$\bar{x}$	s
Children	10	-3.67	33.89
Adults	20	-23.17	50.74

We are interested in testing the hypotheseses $H_0: \sigma_1 = \sigma_2$ and $H_a: \sigma_1 \neq \sigma_2$. The two-sided test statistic is the larger variance divided by the smaller variance and if σ_1 and σ_2 are equal has the $F(n_1 - 1, n_2 - 1)$. Remember that n_1 is the numerator sample size. To organize your calculations, first compute the following three quantities.

$$F = \frac{\text{larger } s^2}{\text{smaller } s^2} =$$

Numerator df $(n_1 - 1) =$

Denominator df $(n_2 - 1) =$

Compare the value of the F you computed to the critical values given in Table D, making sure to go to the row and column corresponding to the appropriate degrees of freedom or as close as you can get to these degrees of freedom. Between which two critical values does F lie? What can be said about the P-value from the table?

COMPLETE SOLUTIONS

Exercise 7.57

Since the adults have the larger standard deviation (and variance), this is the variance that goes into the numerator, so the degrees of freedom are $n_1 - 1 = 20 - 1 = 19$, and $n_2 - 1 = 10 - 1 = 9$.

To compute the value of the test statistic, note that the standard deviations, not the variances, are given as summary statistics. The standard deviations must first be squared to get the two variances. The larger standard deviation is for the adults and the sample variance for this group is $(50.74)^2 = 2574.5476$. For the the children, the sample variance is $(33.89)^2 = 1148.5321$. The value of the test statistic is

$$F = \frac{\text{larger } s^2}{\text{smaller } s^2} = \frac{2574.5476}{1148.5321} = 2.24$$

Since the degrees of freedom (19, 9) are not in the table, we go to the table with the closest values (15, 9) or (20, 9) degrees of freedom. For these, we find that the smallest tabled critical values are 2.34 and 2.30, respectively, corresponding to a significance level of 0.10. The computed value of F is smaller than both critical values. Since the test is two-sided, we double the significance level and conclude that the P-value is greater than 0.20. The data do not show significantly different spreads. Using computer software we find the P-value to be 0.2156.

SELECTED TEXT REVIEW EXERCISES

GUIDED SOLUTIONS

Exercise 7.61

KEY CONCEPTS - matched pairs t test versus two-sample t test, robustness of t procedures

a) Is the matched pairs t test or two-sample t test appropriate here? Did the experimenter match the women in some way?

b) How many degrees of freedom? Use the smaller sample size minus one.

c) What's true about the sample sizes? What, if anything, can you say about outliers or skewness without seeing the original data?

Exercise 7.65

KEY CONCEPTS - one-sample t confidence intervals, checking assumptions

a) With a sample size of only $n = 9$, the most sensible graph would be a stemplot. Complete the stemplot here. Use split stems and use just the numbers to the left of the decimal place.

```
4|
5|
5|
6|
6|
```

b) To compute a level C confidence interval we use the formula $\bar{x} \pm t^* \dfrac{s}{\sqrt{n}}$, where t^* is the upper $(1 - C)/2$ critical value of the $t(n - 1)$ distribution, which can be found in Table C. Fill in the missing values. Don't forget to subtract 1 from the sample size when finding the appropriate degrees of freedom for the t confidence interval.

$$C =$$
$$n =$$
$$t^* =$$

Now compute the values of $\bar{x}$ and s from the data given. Use statistical software or a calculator.

$$\bar{x} = \qquad\qquad s =$$

Substitute all these values into the formula to complete the computation of the 95% confidence interval.

$$\bar{x} \pm t^* \frac{s}{\sqrt{n}} =$$

COMPLETE SOLUTIONS

Exercise 7.61

a) The right test is the two-sample test. The two groups, the control group and the undermine group, consist of 45 women each. However, the 45 women in the two groups are different and there is no indication that the women in the two groups were matched in any way. Thus, there is no natural way to pair observations from the two groups. Such pairing is necessary in a matched pairs t test that uses the differences between matched pairs of observations to compute the t statistic.

b) Both groups have 45 observations. If we use the conservative rule (degrees of freedom are one less than the smaller sample size), we would use 45 - 1 = 44 degrees of freedom.

c) Each group consists of 45 observations. These are large samples (both are greater than 40), so it is safe to use the t procedures, even though the 7-point scale could not have a normal distribution. Note that with a 7-point scale (and sample means of around 4 or 5), the distribution of the observations could not have any extreme outliers and could not be strongly skewed, further reasons why the two sample t procedure should be reasonably accurate.

Exercise 7.65

a) The stemplot follows. There are no outliers and the plot is skewed left. With this few observations, it is difficult to check the assumptions. In this case we might still use the t procedures, but we would not have as much confidence in their validity as we had in most other examples.

```
4|9
5|1 4
5|
6|0 3 3 4 4
6|5
```

b) An approximate 95% confidence interval for the mean percent of nitrogen in ancient air can be calculated from the data on the nine specimens of amber. We use the formula for a t interval, namely $\bar{x} \pm t^* \dfrac{s}{\sqrt{n}}$. In this problem, $\bar{x} = 59.589$, $s = 6.2553$, $n = 9$, hence t^* is the upper $(1 - 0.95)/2 = 0.025$ critical value for the $t(8)$ distribution. From Table C, we see that $t^* = 2.306$. Thus the 95% confidence interval is

$$59.589 \pm 2.306 \frac{6.2553}{\sqrt{9}} = 59.589 \pm 4.808 = (54.78, 64.40)$$

Many statistical software packages will compute a confidence interval for you directly, after you input the data.

CHAPTER 8

INFERENCE FOR PROPORTIONS

SECTION 8.1

OVERVIEW

In this section we consider inference about a population proportion p based on the **sample proportion**

$$\hat{p} = \frac{\text{count of successes in the sample}}{\text{count of observations in the sample}}$$

obtained from an SRS of size n, where X is the number of "successes" (occurrences of the event of interest) in the sample. To use the methods of this section for inference, the following assumptions need to be satisfied.

• The data are an SRS from the population of interest.

• The population is at least 10 times as large as the sample.

• For a test of H_0: $p = p_0$, the sample size n is so large that both np_0 and $n(1-p_0)$ are 10 or more. For a confidence interval, n is so large that both the count of successes $n\hat{p}$ and the count of failures $n(1-\hat{p})$ are 10 or more.

In this case, we can treat $\hat{p}$ as having a distribution that is approximately normal with mean $\mu = p$ and standard deviation $\sigma = \sqrt{p(1-p)/n}$.

An **approximate level C confidence interval** for p is

$$\hat{p} \pm z^* \sqrt{\frac{\hat{p}(1-\hat{p})}{n}}$$

where z^* is the upper $(1-C)/2$ critical value of the standard normal distribution,

$$\sqrt{\frac{\hat{p}(1-\hat{p})}{n}}$$

is the **standard error** of $\hat{p}$, and $z^* \sqrt{\dfrac{\hat{p}(1-\hat{p})}{n}}$ is the **margin of error.**

Tests of the hypothesis $H_0: p = p_0$ are based on the z **statistic**

$$z = \frac{\hat{p} - p_0}{\sqrt{\dfrac{p_0(1-p_0)}{n}}}$$

with P-values calculated from the $N(0, 1)$ distribution.

The **sample size** n required to obtain a confidence interval of approximate margin of error m for a proportion is

$$n = \left(\frac{z^*}{m}\right)^2 p^*(1\text{-}p^*)$$

where p^* is a guessed value for the population proportion and z^* is the upper $(1-C)/2$ critical value of the standard normal distribution. To guarantee that the margin of error of the confidence interval is less than or equal to m no matter what the value of the population proportion may be, use a guessed value of $p^* = 1/2$, which yields

$$n = \left(\frac{z^*}{m}\right)^2$$

GUIDED SOLUTIONS

Exercise 8.3

KEY CONCEPTS - parameters and statistics, proportions

a) To what group does the college president refer?

Population =

Parameter p =

b) A statistic is a number computed from a sample. What is the size of the sample and how many in the sample support firing the coach? From these numbers compute

$$\hat{p} =$$

Exercise 8.7

KEY CONCEPTS - when to use the procedures for inference about a proportion

Recall the assumptions needed to safely use the methods of this section to compute a confidence interval:

- The data are an SRS from the population of interest.

- The population is at least 10 times as large as the sample.

- For a confidence interval, n is so large that both the count of successes $n\hat{p}$ and the count of failures $n(1-\hat{p})$ are 10 or more.

These are the conditions we must check in each of parts a through c. To do so, identify n and compute $\hat{p}$ in each part.

a) n = $\hat{p}$ =

Are the conditions met?

b) $n =$ $\hat{p} =$

Are the conditions met?

c) $n =$ $\hat{p} =$

Are the conditions met? Refer to Example 8.4 in the text for more about the National AIDS Behavioral Surveys.

Exercise 8.13

KEY CONCEPTS - sample size and margin of error

The sample size n required to obtain a confidence interval of approximate margin of error m for a proportion is

$$n = \left(\frac{z^*}{m}\right)^2 p^*(1-p^*)$$

where p^* is a guessed value for the population proportion and z^* is the critical value of the standard normal distribution for the desired level of confidence. To apply this formula here we must determine

$m =$ desired margin of error $=$

$p^* =$ a guessed value for the population proportion $=$

$C =$ desired level of confidence $=$

z^* = the upper $(1-C)/2$ critical value of the standard normal distribution

=

From the statement of the exercise, what are these values? Once you have determined them, use the formula to compute the required sample size n.

$$n = \left(\frac{z^*}{m}\right)^2 p^*(1-p^*) =$$

Exercise 8.15

KEY CONCEPTS - the sampling distribution of a sample proportion, testing hypotheses about a proportion

a) If the appropriate conditions are satisfied, we can treat the sample proportion $\hat{p}$ as having a distribution that is approximately normal with mean $\mu = p$ and standard deviation $\sigma = \sqrt{p(1-p)/n}$. You should check that the appropriate conditions are satisfied, namely,

Is the sample an SRS from a large population?

Is the sample size reasonably large?

Identify p and n.

$p =$ $n =$

The sampling distribution is therefore:

b) What statistical hypotheses should you test to answer this question?

Compute the sample proportion of Harleys stolen in 1995,

$$\hat{p} =$$

Now compute

$$z = \frac{\hat{p} - p_0}{\sqrt{\dfrac{p_0(1 - p_0)}{n}}} =$$

and

$$P\text{-value} =$$

of your test. What do you conclude?

Exercise 8.21

KEY CONCEPTS - testing hypotheses about a proportion, confidence intervals for a proportion

a) First state the hypotheses you will test in terms of p.

Now compute the z-test statistic (identify n and p_0 and calculate $\hat{p}$).

$$z = \frac{\hat{p} - p_0}{\sqrt{\dfrac{p_0(1 - p_0)}{n}}} =$$

For your hypotheses,

P-value =

Is your result significant at the 5% level? (What must the P-value satisfy for this to be true?)

Finally, state your practical conclusions.

b) Recall that an approximate level C confidence interval for p is

$$\hat{p} \pm z^* \sqrt{\frac{\hat{p}(1-\hat{p})}{n}}$$

where z^* is the upper $(1-C)/2$ critical value of the standard normal distribution. From the information given in the problem, provide the following values.

n = sample size =

$\hat{p}$ = sample proportion =

C = level of confidence requested =

z^* = upper $(1-C)/2$ critical value of the standard normal distribution =

Use Table A (or C) to find z^*. Now substitute these values into the formula for the confidence interval to complete the problem.

$$\hat{p} \pm z^* \sqrt{\frac{\hat{p}(1-\hat{p})}{n}} =$$

c) If you are not sure how to answer this, review Section 3.2.

COMPLETE SOLUTIONS

Exercise 8.3

a) The population is presumably all 15,000 living alumni of the college. The parameter p is the proportion of the alumni who support firing the coach.

b) The statistic $\hat{p}$ is the proportion in the SRS that support firing the coach. It has value

$$\hat{p} = 76/200 = 0.38$$

Exercise 8.7

a) $n = 50$ $\qquad\qquad$ $\hat{p} = 14/50 = 0.28$

The data are an SRS from the population of interest.

The population consists of 175 students. This is *not* at least 10 times as large as the sample size $n = 50$.

The methods of this section *cannot* be safely used.

b) $n = 50$ $\qquad\qquad$ $\hat{p} = 38/50 = 0.76$

The data are an SRS from the population of interest.

The population consists of 2400 students. This is at least 10 times as large as the sample size $n = 50$.

$n\hat{p} = 50(38/50) = 38$, $n(1-\hat{p}) = 50(1 - 38/50) = 50(12/50) = 12$. Both of these numbers are at least 10.

The methods of this section *can* be safely used.

c) $n = 2673$ $\qquad\qquad$ $\hat{p} = 0.002$

As discussed in Example 8.4 of the text, the sampling design was a complex stratified sample. The requirement that the sample be an SRS is only approximately met.

The population consists of adult heterosexuals, a number in the millions. This is at least 10 times as large as the sample size $n = 2673$.

$n\hat{p} = 2673(0.002) = 5.346$, $n(1-\hat{p}) = 2673(1 - 0.002) = 2673(0.998) = 2667.654$. The first of these numbers is less than 10.

The methods of this section *cannot* be safely used.

Exercise 8.13

We start with the guess that $p^* = 0.75$. For 95% confidence we use $z^* = 1.96$. The sample size we need for a margin of error $m = 0.04$ is thus

$$n = \left(\frac{z^*}{m}\right)^2 p^*(1 - p^*) = \left(\frac{1.96}{0.04}\right)^2 0.75(1 - 0.75) = 450.1875$$

We round this up to $n = 451$. Thus a sample of size 451 is needed to estimate the proportion of Americans with at least one Italian grandparent who can taste PTC to within ±.04 with 95% confidence.

Exercise 8.15

a) The data are an SRS from the population of interest, all motorcycles stolen in recent years. We must assume that "recent years" includes enough years so that the total number stolen over this period is large (at least 10 times as large as the sample size $n = 9224$).

The sample size is $n = 9224$, which is large.

The appropriate conditions appear to be satisfied. Since $p = 0.14$ and $n = 9224$, we may conclude that the sample proportion $\hat{p}$ has a distribution that is approximately normal with mean $\mu = p = 0.14$ and standard deviation $\sigma = \sqrt{p(1-p)/n} = \sqrt{0.14(1-0.14)/9224} = \sqrt{0.14(0.86)/9224} = 0.0036$.

b) The question being asked can be restated as one in which we wish to test the hypotheses

$$H_0: p = 0.14 \text{ versus. } H_a: p > 0.14$$

The sample proportion for 1995 is

$$\hat{p} = 2490/9224 = 0.27$$

The z-test statistic is therefore

$$z = \frac{\hat{p} - p_0}{\sqrt{\dfrac{p_0(1 - p_0)}{n}}} = \frac{0.27 - 0.14}{\sqrt{\dfrac{0.14(1 - 0.14)}{9224}}} = \frac{0.13}{0.0036} = 36.11$$

This is quite large (from our normal tables), so the P-value of the test is less than 0.0002. We would conclude that there is strong statistical evidence that the proportion of Harleys among stolen bikes is significantly higher than their share of all motorcycles.

Exercise 8.21

a) We wish to see if the majority (more than half) of people prefer the taste of fresh-brewed coffee. The hypotheses to be tested are therefore

$$H_0: p = 0.50 \text{ versus. } H_a: p > 0.50$$

We see that $n = 50$, $p_0 = 0.50$, $\hat{p} = 31/50 = 0.62$, so the z-test statistic is

$$z = \frac{\hat{p} - p_0}{\sqrt{\dfrac{p_0(1 - p_0)}{n}}} = \frac{0.62 - 0.50}{\sqrt{\dfrac{0.50(1 - 0.50)}{50}}} = \frac{0.12}{\sqrt{0.005}} = 1.70$$

The P-value for this value of the z-test statistic is the area to the right of 1.70 under a standard normal curve. From Table A this is

$$P\text{-value} = 0.0446$$

Since this is smaller than 0.05, the result is significant at the 5% level.

We may conclude that there is good statistical evidence that the majority (more than half) of people prefer the taste of fresh-brewed coffee. This conclusion is based on data from 50 subjects of which 62% favored the taste of fresh-brewed coffee.

b) From the information given in the problem,

$$n = \text{sample size} = 50$$

$$\hat{p} = \text{sample proportion} = 0.62$$

$$C = \text{level of confidence requested} = 0.90$$

$z^* =$ upper $(1-C)/2$ critical value of the standard normal distribution
= 1.645

Thus the 90% confidence interval for p is

$$\hat{p} \pm z^* \sqrt{\frac{\hat{p}(1-\hat{p})}{n}} = 0.62 \pm 1.645 \sqrt{\frac{0.62(1-0.62)}{50}}$$

$$= 0.62 \pm 1.645 \sqrt{0.004712}$$

$$= 0.62 \pm 0.113$$

c) You should present the two cups of coffee to the subjects in a random order (i.e., determine which cup a subject gets first by a random mechanism such as flipping a coin).

SECTION 8.2

OVERVIEW

Confidence intervals and tests designed to compare two population proportions are based on the **difference in the sample proportions** $\hat{p}_1$ - $\hat{p}_2$. The formula for the level C confidence interval is

$$(\hat{p}_1 - \hat{p}_2) \pm z^* \text{SE}$$

where z^* is the upper $(1 - C)/2$ standard normal critical value and SE is the standard error for the difference in the two proportions computed as

$$\text{SE} = \sqrt{\frac{\hat{p}_1(1-\hat{p}_1)}{n_1} + \frac{\hat{p}_2(1-\hat{p}_2)}{n_2}}$$

In practice, use this confidence interval when the populations are at least 10 times as large as the samples and the counts of successes and failures are five or more in both samples.

Significance tests for the equality of the two proportions, H_0: $p_1 = p_2$, use a different standard error for the difference in the sample proportions, which is based on a **pooled estimate** of the common (under H_0) value of p_1 and p_2,

$$\hat{p} = \frac{\text{count of successes in both samples combined}}{\text{count of observations in both samples combined}}$$

The test uses the z statistic

$$z = \frac{\hat{p}_1 - \hat{p}_2}{\sqrt{\hat{p}(1 - \hat{p})\left(\dfrac{1}{n_1} + \dfrac{1}{n_2}\right)}}$$

and P-values are computed using Table A of the standard normal distribution. In practice, use this test when the populations are at least 10 times as large as the samples and the counts of successes and failures are five or more in both samples.

GUIDED SOLUTIONS

Exercise 8.25

KEY CONCEPTS - confidence intervals for the difference between two population proportions

Let p_1 represent the proportion of mice ready to breed in good acorn years, p_2 the proportion in bad acorn years. Recall that a level C confidence interval for $p_1 - p_2$ is

$$(\hat{p}_1 - \hat{p}_2) \pm z^*\text{SE}$$

where z^* is the upper $(1 - C)/2$ standard normal critical value and SE is the standard error for the difference in the two proportions computed as

$$\text{SE} = \sqrt{\frac{\hat{p}_1(1 - \hat{p}_1)}{n_1} + \frac{\hat{p}_2(1 - \hat{p}_2)}{n_2}}$$

The two sample sizes are

$n_1 =$

$n_2 =$

From the data, the estimates of these two proportions are

$$\hat{p}_1 =$$

$$\hat{p}_2 =$$

Now compute the standard error

$$SE = \sqrt{\frac{\hat{p}_1(1-\hat{p}_1)}{n_1} + \frac{\hat{p}_2(1-\hat{p}_2)}{n_2}} =$$

For a 90% confidence interval,

$$z^* =$$

Now compute the interval

$$(\hat{p}_1 - \hat{p}_2) \pm z^*SE =$$

Finally, explain why we can use z methods here (under what conditions are z methods applicable?).

Exercise 8.33

KEY CONCEPTS - testing equality of two population proportions, confidence intervals for the difference between two populations proportions

a) Let p_1 represent the proportion of students from urban/suburban backgrounds who succeed and p_2 the proportion from rural/small-town backgrounds who succeed. Recall that a test of the hypothesis $H_0: p_1 = p_2$ uses the z statistic

$$z = \frac{\hat{p}_1 - \hat{p}_2}{\sqrt{\hat{p}(1-\hat{p})\left(\dfrac{1}{n_1} + \dfrac{1}{n_2}\right)}}$$

where n_1 and n_2 are the sizes of the samples, $\hat{p}_1$ and $\hat{p}_2$ the estimates of p_1 and p_2, and

$$\hat{p} = \frac{\text{count of successes in both samples combined}}{\text{count of observations in both samples combined}}$$

First state the hypotheses to be tested (what is the alternative in this case, one-sided or two-sided?).

The two sample sizes are

$n_1 =$

$n_2 =$

From the data, the estimates of these two proportions are

$\hat{p}_1 =$

$\hat{p}_2 =$

Now compute

$$\hat{p} = \frac{\text{count of successes in both samples combined}}{\text{count of observations in both samples combined}} =$$

and then

$$z = \frac{\hat{p}_1 - \hat{p}_2}{\sqrt{\hat{p}(1-\hat{p})\left(\dfrac{1}{n_1} + \dfrac{1}{n_2}\right)}} =$$

Finally, using Table A, compute

    P-value =

What do you conclude?

b) Construct the 90% confidence interval, using the steps outlined to guide you.

The two sample sizes are

    $n_1 =$

    $n_2 =$

From the data, the estimates of these two proportions are

    $\hat{p}_1 =$

    $\hat{p}_2 =$

Now compute

$$SE = \sqrt{\frac{\hat{p}_1(1-\hat{p}_1)}{n_1} + \frac{\hat{p}_2(1-\hat{p}_2)}{n_2}} =$$

For a 90% confidence interval,

$$z^* =$$

Now compute the interval

$$(\hat{p}_1 - \hat{p}_2) \pm z^*\text{SE} =$$

Exercise 8.35

KEY CONCEPTS - testing equality of two population proportions

First state the hypotheses to be tested (what is the alternative in this case; one-sided or two-sided?).

The two sample sizes are

$$n_1 =$$

$$n_2 =$$

From the data, the estimates of these two proportions are

$$\hat{p}_1 =$$

$$\hat{p}_2 =$$

Now compute

$$\hat{p} = \frac{\text{count of successes in both samples combined}}{\text{count of observations in both samples combined}} =$$

and then

$$z = \frac{\hat{p}_1 - \hat{p}_2}{\sqrt{\hat{p}(1-\hat{p})\left(\dfrac{1}{n_1} + \dfrac{1}{n_2}\right)}} =$$

Finally, using Table A, compute

P-value =

What do you conclude?

COMPLETE SOLUTIONS

Exercise 8.25

The two sample sizes are the number of mice trapped in each area:

$$n_1 = 72$$

$$n_2 = 17$$

From the data, the estimates of these two proportions are

$$\hat{p}_1 = 54/72 = 0.75$$

$$\hat{p}_2 = 10/17 = 0.59$$

The standard error is

$$SE = \sqrt{\frac{\hat{p}_1(1-\hat{p}_1)}{n_1} + \frac{\hat{p}_2(1-\hat{p}_2)}{n_2}} = \sqrt{\frac{0.75(1-0.75)}{72} + \frac{0.59(1-0.59)}{17}}$$

$$= \sqrt{0.0026 + 0.0142} = 0.13$$

and for a 90% confidence interval

$$z^* = 1.645$$

so our 90% confidence interval is

$$(\hat{p}_1 - \hat{p}_2) \pm z^*\text{SE} = (0.75 - 0.59) \pm 1.645(0.13) = 0.16 \pm 0.21$$

We can use z methods here since the population of all mice in these forests is most likely much larger than (at least 10 times as large) as the number trapped and since the counts of successes and failures are 54 and 18 for sample 1 (both larger than 5) and 10 and 7 for sample 2 (both larger than 5).

Exercise 8.33

a) We are interested in determining whether there is good evidence that the proportion of students who succeed is *different* for urban/suburban versus rural/small-town backgrounds. Thus the hypotheses to be tested are

$$H_0: p_1 = p_2$$

$$H_a: p_1 \neq p_2$$

The two sample sizes are

$$n_1 = \text{number from urban/suburban background} = 65$$

$$n_2 = \text{number from rural/small-town background} = 55$$

From the data, the estimates of the two proportions of students who succeeded are

$$\hat{p}_1 = 52/65 = 0.8$$

$$\hat{p}_2 = 30/55 = 0.545$$

Next we compute

$$\hat{p} = \frac{\text{count of successes in both samples combined}}{\text{count of observations in both samples combined}}$$

$$= \frac{52 + 30}{65 + 55} = 82/120 = 0.683$$

The value of the z-test statistic is

$$z = \frac{\hat{p}_1 - \hat{p}_2}{\sqrt{\hat{p}(1-\hat{p})\left(\dfrac{1}{n_1} + \dfrac{1}{n_2}\right)}} = \frac{0.8 - 0.545}{\sqrt{0.683(1-0.683)\left(\dfrac{1}{65} + \dfrac{1}{55}\right)}}$$

$$= \frac{0.255}{\sqrt{0.00727}} = 2.99$$

Using Table A (we need to double the tail area since this is a two-sided test),

$$P\text{-value} = 2 \times (0.0014) = 0.0028$$

There is good statistical evidence that the proportion of students who succeed is different for urban/suburban versus rural/small-town backgrounds.

b) The two sample sizes are as in part a, namely,

$$n_1 = 65$$

$$n_2 = 55$$

The estimates of the two proportions are also as in part a, namely,

$$\hat{p}_1 = 52/65 = 0.8$$

$$\hat{p}_2 = 30/55 = 0.545$$

The standard error is

$$SE = \sqrt{\frac{\hat{p}_1(1-\hat{p}_1)}{n_1} + \frac{\hat{p}_2(1-\hat{p}_2)}{n_2}} = \sqrt{\frac{0.8(1-0.8)}{65} + \frac{0.545(1-0.545)}{55}}$$

$$= \sqrt{0.00246 + 0.00451} = 0.08$$

For a 90% confidence interval,

$$z^* = 1.645$$

so our confidence interval is

$$(\hat{p}_1 - \hat{p}_2) \pm z^*\text{SE} = (0.8 - 0.545) \pm 1.645(0.08) = 0.255 \pm 0.13$$

Exercise 8.35

Let p_1 represent the proportion of female students who will succeed and p_2 the proportion of males who will succeed. We are interested in determining whether there is a difference in these two proportions; hence we should test the hypotheses

$$H_0: p_1 = p_2$$

$$H_a: p_1 \neq p_2$$

The two sample sizes are

$$n_1 = \text{number of female students in the course} = 34$$

$$n_2 = \text{number of male students in the course} = 89$$

From the data, the estimates of these two proportions are

$$\hat{p}_1 = 23/34 = 0.6765$$

$$\hat{p}_2 = 60/89 = 0.6742$$

Next we compute

$$\hat{p} = \frac{\text{count of successes in both samples combined}}{\text{count of observations in both samples combined}} = \frac{23 + 60}{34 + 89}$$

$$= 83/123 = 0.6748$$

The value of the z-test statistic is thus

$$z = \frac{\hat{p}_1 - \hat{p}_2}{\sqrt{\hat{p}(1-\hat{p})\left(\dfrac{1}{n_1} + \dfrac{1}{n_2}\right)}} = \frac{0.6765 - 0.6742}{\sqrt{0.6748(1-0.6748)\left(\dfrac{1}{34} + \dfrac{1}{89}\right)}}$$

$$= \frac{0.0023}{\sqrt{0.00892}} = 0.02$$

Finally, we compute the *P*-value using Table A (we need to double the tail area since this is a two-sided test):

$$P\text{-value} = 2 \times (0.4920) = 0.9840$$

The data provide no real statistical evidence of a difference between the proportion of men and women who succeed.

SELECTED TEXT REVIEW EXERCISES

GUIDED SOLUTIONS

Exercise 8.45

KEY CONCEPTS - interpreting confidence intervals

What are the assumptions for inference about a proportion? Review the list given in Section 8.1 of the text. Are any of these assumptions violated?

COMPLETE SOLUTIONS

Exercise 8.45

For the procedures of this chapter to be safe, samples should be simple random samples from the populations of interest. Call-in polls are not simple random samples. You will recall from Chapter 3 that call-in surveys are often biased. Both the value of 81% and the calculations that produce the confidence interval should not be trusted. Fancy statistical procedures do not fix poorly designed surveys.

CHAPTER 9

INFERENCE FOR TWO-WAY TABLES

SECTION 9.1

OVERVIEW

The inference methods of Chapter 8 are first extended to a comparison of more than two population proportions. When comparing more than two proportions, it is best to first do an overall test to see if there is good evidence of any differences among the proportions. Then a detailed follow-up analysis can be performed to decide which proportions are different and to estimate the sizes of the differences.

The overall test for comparing several population proportions arranges the data in a **two-way table**. Two-way tables were introduced in Chapter 2; the tables are a way of displaying the relationship between any two categorical variables. The tables are also called **r x c tables**, where *r* is the number of rows and *c* is the number of columns. Often the rows of the two-way tables correspond to populations or treatment groups, the columns to different categories of the response. When comparing several proportions, there would be only two columns since the response takes only one of two values, and the two columns would represent the successes and failures.

The null hypothesis is $H_0: p_1 = p_2 = \ldots = p_n$, which says that the *n* population proportions are the same. The alternative is "many sided," as the proportions can differ from each other in a variety of ways. To test this hypothesis, we will compare the **observed counts** with the **expected counts** when H_0 is true. The expected cell counts are computed using the formula

$$\text{expected count} = \frac{\text{row total x column total}}{n}$$

where n is the total number of observations.

GUIDED SOLUTIONS

Exercise 9.1

KEY CONCEPTS - $r \times c$ tables, expected counts, comparison of more than two proportions.

a) What are the values of r and c?

$r =$ $c =$

b) To compute the proportion of successful students in each of the three extracurricular activity groups, it is easiest to first fill in the column totals in the table. Note that in this example, the columns correspond to the "populations" and the rows to successes or failures.

	< 2	2 to 12	> 12
C or better	11	68	3
D or F	9	23	5
Totals			

Now, for each of the three extracurricular activity groups, compute the proportion of successful students (C or better).

Activity group	Proportion of successful students
<2 hours	
2 to 12 hours	
>12 hours	

What kind of relationship between extracurricular activities and succeeding in the course is shown by these proportions?

c) Complete the bar chart to compare the three proportions of successes found in part b.

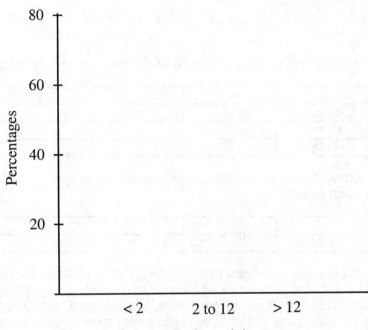

d) Fill in the expected counts in the following table.

	< 2	2 to 12	> 12
C or better	13.78		
D or F			

Remember that the expected cell counts are computed using the formula

$$\text{Expected count} = \frac{\text{row total x column total}}{n}$$

where n is the total number of observations. The value in the upper left-hand corner was obtained as

$$\text{Expected count} = \frac{\text{row total x column total}}{n} = \frac{(82)(20)}{119} = 13.78$$

e) Now compare the observed and expected counts. Are any deviations unusually large? Do the deviations follow the pattern found in part b?

COMPLETE SOLUTIONS

Exercise 9.1

a) There are $r = 2$ rows ("C or better" and "D or F") and $c = 3$ columns ("<2," "2 to 12," and ">12").

b) If we add the counts (students) in each of the three groups we find

	< 2	2 to 12	> 12
C or better	11	68	3
D or F	9	23	5
Totals	20	91	8

From the group (column) totals, we calculate the proportion of successful students in each column.

	< 2	2 to 12	> 12
C or better	11/20 = .550	68/91 = .747	3/8 = .375

The proportions seem to suggest that spending too much time in extracurricular activities is associated with a lack of success in the course. Spending little time in extracurricular activities is more highly associated with success, while spending a moderate amount of time is the group most highly associated with success. Perhaps the maxim "Moderation in all things" applies here!

c) Here is the bar chart of the percentages in part b.

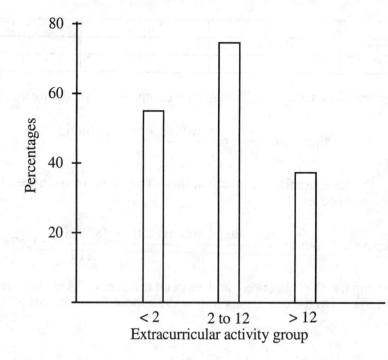

d) Following is the table of expected counts, with the required calculations.

	< 2	2 to 12	> 12
C or better	13.78	62.71	5.51
D or F	6.22	28.29	2.49

$$62.73 = \frac{(82)(91)}{119} \qquad 5.51 = \frac{(82)(8)}{119} \qquad 6.22 = \frac{(37)(20)}{119}$$

$$28.29 = \frac{(37)(91)}{119} \qquad 2.49 = \frac{(37)(8)}{119}$$

e) Here is a table presenting the observed and expected counts together for successful students.

	< 2	2 to 12	> 12
C or better	Observed = 11 Expected = 13.78	Observed = 68 Expected = 62.71	Observed = 3 Expected = 5.51
D or F	Observed = 9 Expected = 6.22	Observed = 23 Expected = 28.29	Observed = 5 Expected = 2.49

None of the deviations is enormous, but neither is there extremely close agreement between the observed and expected counts. The largest deviations between the observed and expected counts (in an absolute sense) occur for the 2 to 12 hours in extracurricular activities group. The observed number of successful students is larger than expected and the observed number of unsuccessful students is smaller than expected. This may suggest an association between success in schoolwork and a moderate involvement in extracurricular activities, as was found in part b.

SECTION 9.2

OVERVIEW

The statistic we will use to compare the expected counts with the observed counts is the **chi-square statistic**. It measures how far the observed and expected counts are from each other using the formula

$$X^2 = \sum \frac{(\text{observed} - \text{expected})^2}{\text{expected}}$$

where we sum up all the $r \times c$ cells in the table.

When the null hypothesis is true, the distribution of the test statistic X^2 is approximately the chi-squared distribution with $(r - 1)(c - 1)$ degrees of freedom. The P-value is the area to the right of X^2 under the chi-square density curve. Use Table E in the back of the book to get the critical values and to compute the P-value. The mean of any chi-squared distribution is equal to its degrees of freedom.

We can use the chi-squared statistic when the data satisfy the following conditions.

- The data are independent SRSs from several populations and each observation is classified according to one categorical variable.
- The data are from a single SRS and each observation is classified according to two categorical variables.
- No more than 20% of the cells in the two-way table have expected counts less than 5.
- All cells have an expected count of at least 1.
- In the special case of the 2 x 2 table, all expected counts should exceed 5.

GUIDED SOLUTIONS

Exercise 9.3

KEY CONCEPTS - computing the X^2 statistic

(a) We reproduce the Minitab output from the text. Expected counts are printed below observed counts.

	<2	2 to 12	>12	Total
A, B, C	11	68	3	82
	13.78	62.71	5.51	
D or F	9	23	5	37
	6.22	28.29	2.49	
Total	20	91	8	119

Chi-Sq = 0.561 + 0.447 + 1.145 + 1.244 + 0.991 + 2.538 = 6.926
DF = 2, P-value = 0.032
1 cells with expected counts less than 5.0

The expected counts were verified in Exercise 9.1. The components of the chi-square statistic are calculated from the expected and observed count in each cell. If you use the expected counts in the table to compute the components of the chi-square statistic, your results will be quite close to but not exactly equal to the values in the Minitab output. This is because the printed output rounds off the expected counts to two decimal places, but the calculations used by Minitab to compute the components of the chi-square statistic are based on "unrounded" values. So if your calculations are based on the "rounded off" values, they will not exactly agree with the values in the Minitab output. Verify the components of the chi-square statistic. We have verified the entry for row 1 and column 1.

Cell in row 1, column 1: $\dfrac{(\text{observed count} - \text{expected count})^2}{\text{expected count}} = \dfrac{(11 - 13.78)^2}{13.78}$

$= 0.561$

Cell in row 1, column 2:

Cell in row 1, column 3:

Cell in row 2, column 1:

Cell in row 2, column 2:

Cell in row 2, column 3:

We calculate the value of the X^2 statistic by summing these 6 components. The sum is

$$X^2 =$$

b) You can read the P-value directly from the output. What is its value, and in simple language, what does it mean to reject H_0 in this setting?

c) Which component of X^2 is the largest? What does this tell you about the relationship between extracurricular activities and academic success?

d) Is this an experiment or an observational study? What would this say about a cause-and-effect relationship?

Exercise 9.5

KEY CONCEPTS - degrees of freedom for the chi-square distribution, P-values using Table E, mean of the chi-square distribution

a) What are the values of r and c in the table? The degrees of freedom can be found using the formula

Degrees of freedom $= (r - 1)(c - 1) =$

b) Go to the row in Table E corresponding to the degrees of freedom found in part a. Between what two entries in Table E does the value $X^2 = 6.926$ lie? What does this tell you about the P-value?

c) What is the relationship between the degrees of freedom and the mean of the X^2 statistic if the null hypothesis were true? How does the observed value of X^2 compare with the mean?

Exercise 9.11

KEY CONCEPTS - two-way tables, testing hypotheses with the X^2 statistic

a) Think of the rows of the table (disease status) as the "treatments" and the columns (olive oil consumption) as the response, since this corresponds to how the samples were selected. To be an experiment, what needs to be true about the assignment of the treatments to the subjects? Was this carried out here?

b) The data from the text are reproduced here, with the expected counts printed below the observed counts. The details of the hand calculations were presented in Exercises 9.1 and 9.3. Verify a few of the expected counts on your own. If you are not sure how to compute the expected counts, review the complete solution for Exercise 9.1.

Olive oil

	Low	Medium	High	Total
Colon cancer	398	397	430	1225
	404.39	404.19	416.42	
Rectal cancer	250	241	237	728
	240.32	240.20	247.47	
Controls	1368	1377	1409	4154
	1371.29	1370.61	1412.10	
Total	2016	2015	2076	6107

Is high olive oil consumption more common among patients without cancer than in patients with colon cancer or rectal cancer? Compute the percentages for each of the three groups to answer this question.

Group	Percentage with high olive oil consumption
Colon cancer	
Rectal cancer	
Controls	

c) We have given you the expected counts. In the formula for the X^2 statistic, remember that there are 9 terms that need to be summed - one for each cell in the table.

$$X^2 = \sum \frac{(\text{observed} - \text{expected})^2}{\text{expected}} =$$

What is the mean of the X^2 statistic under the null hypothesis? If there is evidence to reject the null hypothesis, the computed value of the statistic should be larger than we would expect it to be under the null hypothesis. Is that true in this case?

What is the P-value? What do you conclude?

d) If less than 4% of the cases or controls refused to participate, why would our confidence in the results be strengthened?

COMPLETE SOLUTIONS

Exercise 9.3

a) The components of the chi-square statistic are calculated from the expected and observed count in each cell.

Cell in row 1, column 1: $\dfrac{(\text{observed count - expected count})^2}{\text{expected count}} = \dfrac{(11 - 13.78)^2}{13.78}$

$= 0.561$

Cell in row 1, column 2: $\dfrac{(\text{observed count - expected count})^2}{\text{expected count}} = \dfrac{(68 - 62.71)^2}{62.71}$

$= 0.446$

Cell in row 1, column 3: $\dfrac{(\text{observed count - expected count})^2}{\text{expected count}} = \dfrac{(3 - 5.51)^2}{5.51}$

$= 1.143$

Cell in row 2, column 1: $\dfrac{(\text{observed count - expected count})^2}{\text{expected count}} = \dfrac{(9 - 6.22)^2}{6.22}$

$= 1.243$

Cell in row 2, column 2: $\dfrac{(\text{observed count - expected count})^2}{\text{expected count}} = \dfrac{(23 - 28.29)^2}{28.29}$

$= 0.989$

Cell in row 2, column 3: $\dfrac{(\text{observed count - expected count})^2}{\text{expected count}} = \dfrac{(5 - 2.49)^2}{2.49}$

$= 2.530$

We calculate the value of the X^2 statistic by summing these 6 components. The sum is

$$X^2 = 6.912$$

b) The *P*-value is the probability of obtaining a value of the X^2 statistic at least as large as the observed value (6.926 on the Minitab output) and is given in the output as 0.031. We would reject H_0 at level 0.05 but not at level 0.01. Rejecting H_0 means that the observed differences in the proportions in each group for the successful ("C or better") and unsuccessful students ("D or F") cannot be easily attributed to chance. In other words, there is evidence that the proportions of successful students in the three groups (representing amounts of time spent on extracurricular activities) are different.

c) The term contributing most to X^2 is the term for the number of unsuccessful students spending more than 12 hours per week on extracurricular activities. This points to the fact that if one spends too much time on extracurricular activities, one has little time for schoolwork and is thus likely to be unsuccessful.

d) The study was not a designed experiment and hence does not prove that spending more or less time on extracurricular activity *causes* changes in academic success. It is an observational study. While at first glance it seems plausible that changing the amount of time spent on extracurricular activities ought to cause changes in academic success, this assumes (among other things) that students will trade time spent on schoolwork with time spent on extracurricular activities. However, some students may seek out involvement in extracurricular activities as an escape from schoolwork. If they aren't involved in extracurricular activities, they will simply "goof off." Changes in extracurricular activity thus will not necessarily produce changes in academic performance.

Exercise 9.5

a) As we saw in Exercise 9.1, there are $r = 2$ rows ("C or better" and "D or F") and $c = 3$ columns ("<2," "2 to 12", and ">12"). Thus the number of degrees of freedom is

$$(r - 1)(c - 1) = (2 - 1)(3 - 1) = (1)(2) = 2$$

which agrees with the value in the Minitab output (see Exercise 9.3).

b) If we look in the df = 2 row of Table E, we find the following information.

p	.05	.025
x^*	5.99	7.38

$X^2 = 6.926$ lies between the entries for $p = .05$ and $p = .025$. This tells us that the P-value is between .05 and .025.

c) The mean of the X^2 statistic if the null hypothesis is true is equal to the degrees of freedom, which is equal to 2 in this case. The observed value is considerably larger than the mean (what we would expect if H_0 were true). This is why there is evidence that the null hypothesis is not true - the differences between the observed and expected counts as measured by X^2 are considerably larger than we would expect to find if the null hypothesis were true.

Exercise 9.11

a) We are thinking of the disease groups as the explanatory variable, since the samples were chosen from these groups. The experimenters did not assign the subjects to the disease groups. This makes the data observational, and there is the possibility that differences in olive oil consumption may be a result of confounding variables, not necessarily disease status. (It might be more natural to think of olive oil consumption as the explanatory variable and disease status as the response, but the samples were not selected from the different olive oil consumption groups.)

b) The percentages in each group with high olive oil consumptions are computed in the table. There appears to be very little difference in these percentages, suggesting that olive oil in the diet may be unrelated to these forms of cancer.

Group	Percentage with high olive oil consumption
Colon cancer	430/1225 = 0.351
Rectal cancer	237/728 = 0.326
Controls	1409/4154 = 0.339

c) From the computer output,

$$X^2 = \sum \frac{(\text{observed} - \text{expected})^2}{\text{expected}} = 0.101 + 0.128 + 0.443 +$$
$$0.390 + 0.003 + 0.443 +$$
$$0.008 + 0.030 + 0.007 = 1.552$$

If you used the expected counts in the table, which were rounded to two decimals, your answers may disagree slightly with the computations due to rounding error. The mean of the X^2 statistic is equal to the degrees of freedom, which in this case is $(r - 1)(c - 1) = (3 - 1)(3 - 1) = 4$. Since the numerical value of the X^2 statistic is below the mean, there is little evidence to reject the null hypothesis. As can be seen from the table, the observed and expected counts are quite close. From Table E, you can see that the P-value is greater than 0.25. In fact, using statistical software shows the P-value = 0.817. These data show no evidence of a relationship between disease and olive oil consumption.

d) If there is a high *nonresponse rate*, then it is possible for the percentages in the table to be quite different than the true percentages due to response bias. In this case, the existence of or lack of a relationship could potentially be due to this bias. However, with 96% of the people responding, the bias if it exists could not be large and would have little impact on the results.

CHAPTER 10

ONE-WAY ANALYSIS OF VARIANCE: COMPARING SEVERAL MEANS

SECTION 10.1

OVERVIEW

The two-sample t procedures compare the means of two populations. However, when the mean is the best description of the center of a distribution, we may want to compare several population means or several treatment means in a designed experiment. For example, we might be interested in comparing the mean weight loss by dieters on three different diet programs or the mean yield of four varieties of green beans.

The method we use to compare more than two population means is the **analysis of variance (ANOVA) F test**. This test is also called the **one-way ANOVA**. The ANOVA F test is an overall test that looks for any difference between a group of I means. The null hypothesis is H_0: $\mu_1 = \mu_2 = \ldots = \mu_I$, where we tell the population means apart by using the subscripts 1 though I. The alternative hypothesis is H_a: not all the means are equal. In a more advanced course, you would study formal inference procedures for a follow-up analysis to decide which means differ and to estimate how large the differences are. Note that formally the ANOVA F test is a different test from the F test you may have studied in Chapter 7 that compared the standard deviations of two populations, although the ANOVA F test does involve the comparison of two measures of variation.

The ANOVA F test compares the variation among the groups to the variation within the groups through the **F statistic**,

$$F = \frac{\text{variation among the sample means}}{\text{variation among the individuals in the same sample}}$$

The important thing to take away from this chapter is the rationale behind the ANOVA F test. The particulars of the calculation are not as important since software usually calculates the numbers for us.

The F statistic has the F distribution. The distribution is completely defined by its two degrees of freedom parameters, the numerator degrees of freedom and the denominator degrees of freedom. The numerator has $I - 1$ degrees of freedom, where I is the number of populations we are comparing. The denominator has $N - I$ degrees of freedom, where N is the total number of observations. The F distribution is usually written $F(I - 1, N - I)$.

The assumptions for ANOVA are that

- There are I independent SRSs.
- Each population is normally distributed with its own mean, μ_i.
- All populations have the same standard deviation, σ.

The first assumption is the most important. The test is robust against nonnormality, but it is still important to check for outliers and/or skewness that would make the mean a poor measure of the center of the distribution. As for the assumption of equal standard deviations, make sure that the largest sample standard deviation is no more than twice the smallest standard deviation.

GUIDED SOLUTIONS

Exercise 10.1

KEY CONCEPTS - stemplots, comparing means, interpreting ANOVA output

a) Complete the stemplots (they use split stems). Don't forget to round to the nearest whole number of beats. From the stemplots, would you say that any of the groups show outliers or extreme skewness? You may want to review the material on stemplots in Chapter 1 if you have forgotten the details.

Control	With pet	With friend
5	5	5
6	6	6
6	6	6
7	7	7
7	7	7
8	8	8
8	8	8
9	9	9
9	9	9
10	10	10

b) A comparison of the three means is interesting. We have used the same "axes" for the stemplots, so the relationship between the means is quite clear from the plots. Does the presence of a pet or a friend reduce heart rate during a stressful task?

c) From the output, determine the values of the ANOVA F statistic and its P-value.

F statistic = P-value =

What do you conclude? You might want to look at the confidence intervals for the individual means as well. Is anything surprising about these data?

Exercise 10.3

KEY CONCEPTS - ANOVA degrees of freedom, computing P-values from Table D

a) In the table, fill in the numerical values and explain in words the meaning of each symbol we are using in the notation for the one-way ANOVA. For consistency, let group 1 be the control, group 2 be the "With pet" group and group 3 be the "With friend" group.

Symbol	Value	Verbal meaning
I		
n_1		
n_2		
n_3		
N		

b) Use the text formulas and the results from part a to give the numerator and denominator degrees of freedom.

Numerator degrees of freedom =

Denominator degrees of freedom =

c) The value $F = 14.08$ needs to be referred to an $F(2, 42)$ distribution. Use denominator degrees of freedom equal to 40 to be conservative, since the entry for 42 is not in the table. What can you say about the P-value from Table D?

Exercise 10.7

KEY CONCEPTS - comparison between means, checking assumptions, interpreting results from an ANOVA

a) By comparing the group means, what is the relationship between marital status and salary?

b) The ratio of the largest to the smallest standard deviations is

$$\frac{\text{largest sample standard deviation}}{\text{smallest sample standard deviation}} =$$

Does this allow the use of the ANOVA F test?

c) Calculate the degrees of freedom for the ANOVA F test by first computing N and I.

$N =$ $\qquad\qquad\qquad$ $I =$

$\qquad$ Numerator df = $\qquad\qquad\qquad$ Denominator df =

d) The large sample sizes (particularly for the married men) indicate that the margins of error for the sample means will be very small, much smaller than the observed differences in the means. What is the numerical value of the standard error for married men? How does it compare to the differences in mean salaries between married men and the other groups?

e) Is this an observational study or an experiment? How does that affect the type of conclusions that can be made?

COMPLETE SOLUTIONS

Exercise 10.1
a) The following are completed stemplots . There isn't extreme skewness in any of the plots. The "With pet" group has one high observation and the control group has one low observation. But neither is so far removed from the rest of the data that it will seriously affect the procedures.

Control	With pet	With friend
5|	5|9	5|
6|3	6|4	6|
6|	6|5 9 9 9	6|
7|1 3	7|0 0 0 2	7|
7|5 8	7|6	7|7
8|0	8|0	8|0 2 3
8|5 5 5 7 7 8	8|5 6	8|7 8
9|0 2	9|	9|0 1 2
9|9	9|8	9|7 8
10|	10|	10|0 1 1 2

b) The mean heart rates are control = 82.524, "With friend" = 91.325, and "With pet" = 73.483. The mean heart rate is lowest for those who did the stressful task with their pet. Surprisingly, the mean heart rate was higher for those with a good friend present than for those who were alone.

c) Reading from the computer output, the F statistic = 14.08 and its P-value = 0.000. There is very strong evidence of some difference between the three groups. While we are not going to do a "formal" follow-up analysis, the confidence intervals for the means barely overlap, suggesting differences among all three groups. The surprising thing about the data is that the presence of a friend not only doesn't reduce stress, it appears to make the situation worse.

Exercise 10.3
a) The table gives the value and states in words the meaning of each symbol used in a one-way ANOVA.

Symbol	Value	Verbal meaning
I	3	Number of groups
n_1	15	Number of women in the control group
n_2	15	Number of women in the "With pet" group
n_3	15	Number of women in the "With friend" group
N	45	Total number of women in the experiment

b) The ANOVA F statistic has the F distribution with $I - 1 = 3 - 1 = 2$ degrees of freedom in the numberator and $N - I = 45 - 3 = 42$ degrees of freedom in the denominator.

c) The critical value of 8.25 corresponds to a tail probability of 0.001 for an $F(2, 40)$ distribution. Since the value $F = 14.08$ exceeds this, we can say that the P-value is less than 0.001, which agrees with the computer output.

Exercise 10.7
a) The most obvious feature is that men who are or have been married earn more, on average, than single men. Men who are or have been married earn about the same amount, although divorced men appear to earn a little less, on average, then married or widowed men.

b) The ratio of the largest to the smallest standard deviations is

$$\frac{\text{largest sample standard deviation}}{\text{smallest sample standard deviation}} = \frac{8119}{5731} = 1.42 < 2$$

Since this ratio is less than 2, the sample standard deviations allow the use of the ANOVA F test.

c) We calculate

I = number of populations we wish to compare
= number of different marital statuses
= 4

n_1 = sample size from population 1 (single men) = 337
n_2 = sample size from population 2 (married men) = 7730
n_3 = sample size from population 3 (divorced men) = 126
n_4 = sample size from population 4 (widowed men) = 42

N = total sample size = sum of the n_i = 8235.

The degrees of freedom for the ANOVA F statistic are

$$I - 1 = 4 - 1 = 3$$

degrees of freedom in the numerator and

$$N - I = 8235 - 4 = 8231$$

degrees of freedom in the denominator.

d) The large sample sizes (particularly for the married men) indicate that the margins of error for the sample means will be very small, much smaller than the observed differences in the means. For example, the standard error for the mean of married men is

$$\frac{\$7159}{\sqrt{7730}} = \$81.4$$

which is very small compared with the difference of, for example, over $5000 in mean salaries with single men (you can check that the standard error for single men is $312.2, also much smaller than the difference in means of over $5000).

e) The differences in means probably do not mean that getting married raises men's mean incomes. This is an observational study, so it is not safe to conclude that the observed differences are due to cause and effect. A likely explanation of the observed differences is that the typical single man is much younger than the typical married man. As men get older they are more likely to be married. Younger men have been in the firm for less time than older men and so will have lower salaries. This explanation may also explain the small differences between men who are or have been married. Married and widowed men (as compared to divorced men) are likely to include the most senior men in the firm. The most senior men are likely to have the highest salaries.

SECTION 10.2

OVERVIEW

Although it is generally best to leave the ANOVA computations to statistical software, seeing the formulas sometimes helps one to obtain a better understanding of the procedure. In addition, there are times when the original data are not available, and you have only the group means and standard deviations or standard error. In these instances, the formulas described here are required to carry out the ANOVA F test.

The F statistic is $F = \dfrac{\text{MSG}}{\text{MSE}}$, where MSG is the **mean square for groups**,

$$\text{MSG} = \frac{n_1(\bar{x}_1 - \bar{x})^2 + n_2(\bar{x}_2 - \bar{x})^2 + \ldots + n_I(\bar{x}_I - \bar{x})^2}{I - 1}$$

with

$$\bar{x} = \frac{n_1\bar{x}_1 + n_2\bar{x}_2 + \ldots + n_I\bar{x}_I}{N}$$

and MSE is the **error mean square**,

$$\text{MSE} = \frac{s_1^2(n_1 - 1) + s_2^2(n_2 - 1) + \ldots + s_I^2(n_I - 1)}{N - I}.$$

Because MSE is an average of the individual sample variances, it is also called the **pooled sample variance**, written s_P^2, and its square root, $s_p = \sqrt{\text{MSE}}$ is called the **pooled standard deviation**.

We can also make a confidence interval for any of the means by using the formula $\bar{x}_i \pm t * \dfrac{s_p}{\sqrt{n_i}}$. The critical value is $t*$ from the t distribution with $N - I$ degrees of freedom.

GUIDED SOLUTIONS

Exercise 10.9

KEY CONCEPTS - ANOVA computations, standard errors

a) The data are given in the form "mean ± standard error." The computation of the ANOVA F statistic requires the three group means and standard deviations. The means are given directly, and the standard deviations can be derived easily from the standard error (SE) by recalling that

$$SE = \frac{s}{\sqrt{n}}, \text{ or equivalently } s = SE\sqrt{n}$$

To simplify the calculations that need to be done, complete the following table. We have completed the first line for you, where we are given that $n_1 = 16$ eggs hatched at the cold temperature, $\bar{x}_1 = 28.89$, $SE_1 = 8.08$, and we compute

$$s_1 = SE_1 \sqrt{n_1} = 8.08\sqrt{16} = 32.32$$

Temperature	$\bar{x}_i$	SE_i	n_i	s_i
Cold	28.89	8.08	16	32.32
Neutral				
Hot				

Do the standard deviations satisfy the rule of thumb for using ANOVA?

$$\frac{\text{largest sample standard deviation}}{\text{smallest sample standard deviation}} =$$

b) You will need the means, sample sizes and standard deviations from your table in part a to do the calculations. To compute MSG, you first need to compute the overall mean

$$\bar{x} = \frac{n_1\bar{x}_1 + n_2\bar{x}_2 + ... + n_I\bar{x}_I}{N} =$$

and then substitute the means, sample sizes, and overall mean into the formula

$$MSG = \frac{n_1(\bar{x}_1 - \bar{x})^2 + n_2(\bar{x}_2 - \bar{x})^2 + \ldots + n_I(\bar{x}_I - \bar{x})^2}{I-1} =$$

MSE is then obtained from the formula

$$MSE = \frac{s_1^2(n_1 - 1) + s_2^2(n_2 - 1) + \ldots + s_I^2(n_I - 1)}{N - I} =$$

Finally,

$$F = \frac{MSG}{MSE} =$$

What are the degrees of freedom for the ANOVA F statistic? Compare the value you calculated to the critical values in Table D. What is the P-value and is there evidence that nest temperature affects the mean weight of newly hatched pythons?

COMPLETE SOLUTIONS

Exercise 10.9
a) The completed table follows.

Temperature	$\bar{x}_i$	SE_i	n_i	s_i
Cold	28.89	8.08	16	32.320
Neutral	32.93	5.61	38	34.582
Hot	32.27	4.10	75	35.507

The ratio of the largest to the smallest standard deviations is

$$\frac{\text{largest sample standard deviation}}{\text{smallest sample standard deviation}} = \frac{35.507}{32.320} = 1.10 < 2$$

so the rule of thumb is satisfied.

b) Using the means, sample sizes, and standard deviations from the table in part a

$$\bar{x} = \frac{n_1\bar{x}_1 + n_2\bar{x}_2 + \ldots + n_I\bar{x}_I}{N} = \frac{(16)(28.89) +)(38)(32.93) + (75)(32.27)}{129} = 32.045.$$

since $N = 16 + 75 + 38 = 129$.

Substituting into the formula for MSG,

$$\text{MSG} = \frac{n_1(\bar{x}_1 - \bar{x})^2 + n_2(\bar{x}_2 - \bar{x})^2 + \ldots + n_I(\bar{x}_I - \bar{x})^2}{I - 1} =$$

$$= \frac{16(28.89 - 32.045)^2 + 38(32.93 - 32.045)^2 + 75(32.27 - 32.045)^2}{3 - 1}$$

$$= 96.412$$

To complete the calculations, we have

$$\text{MSE} = \frac{s_1^2(n_1 - 1) + s_2^2(n_2 - 1) + \ldots + s_I^2(n_I - 1)}{N - I}$$

$$= \frac{32.320^2(15) + 34.582^2(37) + 35.507^2(74)}{126} = 1215.975$$

Finally,

$$F = \frac{\text{MSG}}{\text{MSE}} = \frac{96.412}{1215.975} = 0.0793$$

The F-value should be compared to critical values from the $F(2, 126)$ distribution or, being conservative, to the $F(2, 100)$ distribution. The 0.10 critical value is 2.36, so there is no statistical evidence that nest temperature affects the mean weight of newly hatched pythons.

SELECTED TEXT REVIEW EXERCISES

Exercise 10.15

KEY CONCEPTS - side-by-side stemplots, ANOVA assumptions, drawing conclusions from ANOVA output

a) Complete the stemplots (they use split stems). From the stemplots, would you say that any of the groups show outliers or extreme skewness?

Never logged	Logged 1 year ago	Logged 8 years ago
0	0	0
0	0	0
1	1	1
1	1	1
2	2	2
2	2	2
3	3	3

b) Do the standard deviations satisfy the rule of thumb for using ANOVA?

$$\frac{\text{largest sample standard deviation}}{\text{smallest sample standard deviation}} =$$

What do the means suggest about the effect of logging? Based on the F statistic and P-value, state your overall conclusions.

COMPLETE SOLUTION

Exercise 10.15
a) The side-by-side stemplots are completed here. Extreme skewness is not evident. There is a low outlier in the "Logged 8 years ago" column. The analysis was done with and without this observation, and it was found to have little effect on the group mean or the overall one-way ANOVA. If you are

unclear about whether an outlier can affect the results, the analysis can be done with and without these points.

Never logged	Logged 1 year ago	Logged 8 years ago
0		0	2	0	4
0		0	9	0	
1		1	2 2 4 4	1	2 2
1	6 9 9	1	5 7 7 8 9	1	5 8 8 9
2	0 1 2 4	2	0	2	2 2
2	7 7 8 9	2		2	
3	3	3		3	

b) The ratio of the largest to the smallest standard deviations is

$$\frac{\text{largest sample standard deviation}}{\text{smallest sample standard deviation}} = \frac{5.761}{4.981} = 1.16 < 2$$

so the standard deviations satisfy the rule of thumb for safe use of ANOVA.

The overall ANOVA F test has $F = 11.43$ and P-value $= 0.000$, so there is strong evidence of a difference in mean trees per plot among the three groups. Examination of the means shows both the logged groups to have lower mean trees per plot than the never logged group, but there appears to be little difference between the plots logged 1 year ago and the plots logged 8 years ago. This is supported by the confidence intervals for the three means, which show no overlap between the interval for the never logged group and the other two intervals but considerable overlap between the confidence intervals for the two logged groups.

CHAPTER 11

INFERENCE FOR REGRESSION

SECTIONS 11.1 and 11.2

OVERVIEW

In Chapter 2 we first encountered regression. The assumptions that describe the regression model we use in this chapter are the following.

- We have n observations on an explanatory variable x and a response variable y. Our goal is to study or predict the behavior of y for given values of x.

- For any fixed value of x, the response y varies according to a normal distribution. Repeated responses y are independent of each other.

- The mean response μ_y has a straight-line relationship with x:

$$\mu_y = \alpha + \beta x$$

The slope β and intercept α are unknown parameters.

- The standard deviation of y (call it σ) is the same for all values of x. The value of σ is unknown.

The **true regression line** is $\mu_y = \alpha + \beta x$ and says that the mean response μ_y moves along a straight line as the explanatory variable x changes. The parameters β and α are estimated by the slope b and intercept a of the least-squares regression line, and the formulas for these estimates are

$$b = r\frac{s_y}{s_x}$$

and

$$a = \bar{y} - b\bar{x}$$

where r is the correlation between y and x, $\bar{y}$ is the mean of the y observations, s_y is the standard deviation of the y observations, $\bar{x}$ is the mean of the x observations, and s_x is the standard deviation of the x observations.

The **standard error about the least-squares line** is

$$s = \sqrt{\frac{1}{n-2}\sum \text{residual}^2} = \sqrt{\frac{1}{n-2}\sum (y - \hat{y})^2}$$

where $\hat{y} = a + bx$ is the value we would predict for the response variable based on the least-squares regression line. We use s to estimate the unknown σ in the regression model.

A **level C confidence interval** for β is

$$b \pm t^*SE_b$$

where t^* is the upper $(1 - C)/2$ critical value for the t distribution with $n - 2$ degrees of freedom and

$$SE_b = \frac{s}{\sqrt{\sum (x - \bar{x})^2}}$$

is the standard error of the least-squares slope b. SE_b is usually computed using a calculator or statistical software.

The **test of the hypothesis** $H_0: \beta = 0$ is based on the t statistic

$$t = \frac{b}{SE_b}$$

with *P*-values computed from the *t* distribution with *n* - 2 degrees of freedom. This test is also a test of the hypothesis that the correlation is 0 in the population.

A **level *C* confidence interval for the mean response** μ_y when *x* takes the value *x** is

$$\hat{y} \pm t^* \, \mathrm{SE}_{\hat{\mu}}$$

where $\hat{y} = a + bx$, *t** is the upper (1 - *C*)/2 critical value for the *t* distribution with *n* - 2 degrees of freedom and

$$\mathrm{SE}_{\hat{\mu}} = s \sqrt{\frac{1}{n} + \frac{(x^* - \bar{x})^2}{\sum (x - \bar{x})^2}}$$

$\mathrm{SE}_{\hat{\mu}}$ is usually computed using a calculator or statistical software.

A **level *C* prediction interval for a single observation** on *y* when *x* takes the value *x** is

$$\hat{y} \pm t^* \, \mathrm{SE}_{\hat{y}}$$

where *t** is the upper (1 - *C*)/2 critical value for the *t* distribution with *n* - 2 degrees of freedom and

$$\mathrm{SE}_{\hat{y}} = s \sqrt{1 + \frac{1}{n} + \frac{(x^* - \bar{x})^2}{\sum (x - \bar{x})^2}} \, .$$

$\mathrm{SE}_{\hat{y}}$ is usually computed using a calculator or statistical software.

Finally, it is always good practice to check that the data satisfy the linear regression model assumptions before doing inference. Scatterplots and residual plots are useful tools for checking these assumptions.

GUIDED SOLUTIONS

Exercise 11.1

KEY CONCEPTS - scatterplots, correlation, linear regression, residuals, standard error of the least-squares line

a) Sketch your scatterplot on the axes provided below.

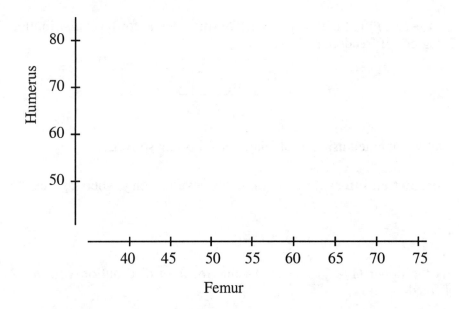

Now use your calculator (or statistical software) to compute the correlation r and the equation of the least-squares regression line.

$r =$

Humerus length =

Do you think femur length will allow a good prediction of humerus length?

b) What does the slope β of the true regression line say about *Archaeopteryx*?

Enter your estimates of the slope β and intercept α of the true regression line in the space provided. Refer to your answer in part a for these estimates.

Estimate of $\beta =$

Estimate of $\alpha =$

c) To compute the residuals, complete the table.

Observed value of humerus length	Predicted value of humerus length : -3.65959 + 1.19690(femur length)	Residual (observed - predicted length)
41		
63		
70		
72		
84		
	Sum =	

Now estimate the standard deviation σ by computing

$$\sum \text{residual}^2 =$$

and then completing the following.

$$s = \sqrt{\frac{1}{n-2} \sum \text{residual}^2} =$$

Exercise 11.13

KEY CONCEPTS - scatterplots, outliers and influential observations, r^2, confidence intervals for the slope

a) Use the axes provided to make your scatterplot.

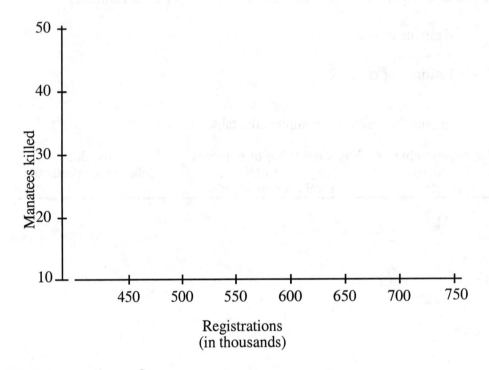

What do you observe?

b) Refer to Section 2.3 if you need to review the meaning of r^2.

c) Explain what the slope β means in this setting.

To determine the confidence interval, recall that a level C confidence interval for β is

$$b \pm t^* SE_b$$

where t^* is the upper $(1 - C)/2$ critical value for the t distribution with $n - 2$ degrees of freedom and

$$SE_b = \frac{s}{\sqrt{\sum(x - \bar{x})^2}}$$

is the standard error of the least-squares slope b. In this example, b and SE_b can be read directly from the output given in part b (look in the row labeled by the explanatory variable). Their values are

$b =$

$SE_b =$

Now find t^* for a 90% confidence interval from Table C (what is n here?).

$t^* =$

Put all these pieces together to compute the 90% confidence interval.

$b \pm t^* SE_b =$

Exercise 11.14

KEY CONCEPTS - prediction, prediction intervals

a) Use the output given in Exercise 11.13 to determine the equation of the least-squares regression line.

Manatees killed =

Now use this equation to predict the number of manatees that will be killed in a year when 716,000 powerboats are registered. Remember, boats were measured in thousands of boats, so you should substitute 716, not 716,000, into the equation of the least-squares regression line.

Manatees killed =

b) Compare your answer in part a with the prediction of 47.97. If they differ, can you explain why?

Two 95% intervals are given in the output. Which of these is appropriate for predicting the number killed in a future year?

c) Compare the number killed in 1991, 1992, and 1993 with the 95% prediction interval in part b.

Exercise 11.15

KEY CONCEPTS - confidence intervals for the mean response

A level C confidence interval for the mean response μ_y when x takes the value $x*$ is

$$\hat{y} \pm t* \, \text{SE}_{\hat{\mu}}$$

where $\hat{y} = a + bx*$, $t*$ is the upper $(1 - C)/2$ critical value for the t distribution with $n - 2$ degrees of freedom, and

$$\text{SE}_{\hat{\mu}} = s \sqrt{\frac{1}{n} + \frac{(x*-\bar{x})^2}{\sum(x-\bar{x})^2}}$$

In this case, we are given $\text{SE}_{\hat{\mu}}$ and do not need to compute it from the formula. What is the value of $\text{SE}_{\hat{\mu}}$?

$$\text{SE}_{\hat{\mu}} =$$

Next, complete the following. Note that $\hat{y}$ was computed in Exercise 11.14.

$\hat{y}$ = predicted number of manatees killed for 716,000 powerboats

$\qquad$ =

$n - 2 =$

t^* for a 90% confidence interval (refer to Table C) =

The desired 90% confidence interval is therefore

$\hat{y} \pm t^* \, \mathrm{SE}_{\hat{\mu}} =$

Exercise 11.18

KEY CONCEPTS - scatterplots, tests for the slope of the least-squares regression line, prediction, prediction intervals

a) Use the axes to draw your scatterplot. We have placed % return-overseas on the vertical axis since this is the response variable.

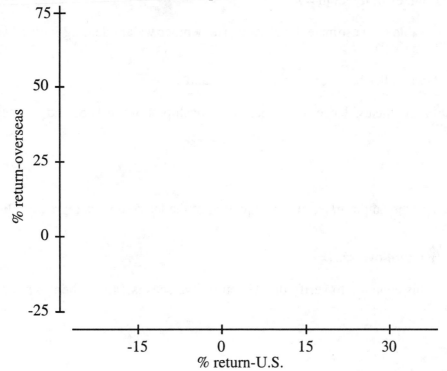

b) The test of the hypothesis $H_0: \beta = 0$ is based on the t statistic

$$t = \frac{b}{\text{SE}_b}$$

with P-values computed from the t distribution with n - 2 degrees of freedom. Both b and SE_b can be read from the Minitab output given in the problem (look at Example 11.4 in the text for how to read this output). Write their values.

$b =$

$\text{SE}_b =$

Now compute t.

$$t = \frac{b}{\text{SE}_b} =$$

What are the degrees of freedom for t? Refer to the original data in Chapter 2 to determine the sample size n.

Degrees of freedom = n - 2 =

Now use Table C to estimate the P-value (remember we are dealing with a two-sided alternative!)

P-value lies between   and

How strong is the evidence for a linear relationship between U.S. and overseas returns?

c) Based on the output in part b, the equation of the least-squares regression line is

% return-overseas =

Now use this equation to verify that % return-overseas is 14.95 when % return-U.S. is 15.

Two 90% intervals are given in the output. Which is the appropriate interval for the return on foreign stocks next year if U.S. stocks return 15%?

d) Is the regression prediction useful in practice? Use the r^2 value to help explain your answer.

Exercise 11.19

KEY CONCEPTS - examining residuals

a) Use the axes to make your residual plot.

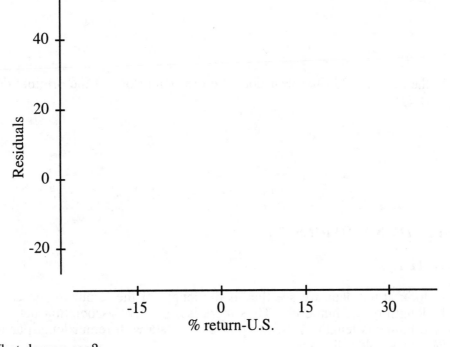

What do you see?

b) Make either a stemplot or histogram of the residuals.

Describe the shape of your plot. In what way is it somewhat nonnormal?

Identify the outlier. To what year does it correspond (look at the original data in Chapter 2)?

COMPLETE SOLUTIONS

Exercise 11.1

a) If we look at the data, we see that as the length of the femur increases, so does the length of the humerus. Thus there is a positive association between femur and humerus length. A scatterplot of the data with femur length as the explanatory variable follows.

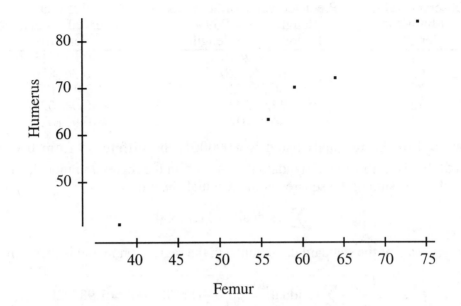

The scatterplot indicates a fairly strong positive association between femur and humerus length. If we calculate the correlation r and the equation of the least-squares line, we obtain the following.

$$r = 0.994$$

$$\text{Humerus length} = -3.65959 + 1.19690 \text{ (femur length)}$$

The correlation is very high, so one would expect that femur length would allow good prediction of humerus length.

b) The slope β of the true regression line tells us the mean increase (in cm) in the length of the humerus associated with a 1-cm increase in the length of the femur in *Archaeopteryx*. From the data,

$$\text{Estimate of } \beta = 1.19690$$

the slope of the least-squares regression line. From the data,

$$\text{Estimate of } \alpha = -3.65959$$

the intercept of the least-squares regression line.

c) The residuals for the five data points are given in the table.

Observed value of humerus length	Predicted value of humerus length : -3.65959 + 1.19690(femur length)	Residual (observed - predicted length)
41	41.822618	-0.822618
63	63.366820	-0.366820
70	66.957520	3.042480
72	72.942021	-0.942021
84	84.911022	-0.911022

The sum of the residuals listed is -.000001, the difference from 0 due to roundoff. To estimate the standard deviation σ in the regression model, we first calculate the sum of the squares of the residuals listed:

$$\sum \text{residual}^2 = 11.785306$$

Our estimate of the standard deviation σ in the regression model is therefore

$$s = \sqrt{\frac{1}{n-2}\sum \text{residual}^2} = \sqrt{\frac{1}{5-2}(11.785306)} = 1.982028$$

Exercise 11.13

a) Here is a scatterplot showing the relationship between power boats registered and manatees killed. Since we are trying to use the number of power boat registrations to explain the number of manatees killed, power boat registrations is the explanatory variable.

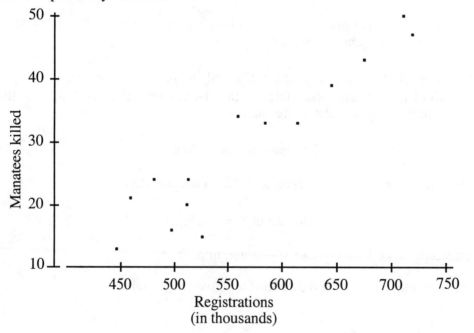

The overall pattern is roughly linear with a positive slope. There are no clear outliers or strongly influential data points.

b) We recall that r^2 tells us the percentage of the variation in the response variable accounted for by the explanatory variable in the least-squares regression line. In this case $r^2 = 0.886$; thus 88.6% of the variation in manatees killed is accounted for by power boat registrations.

c) The slope β of the true regression line tells us the mean increase in the manatee deaths associated with an increase of 1000 power boat registrations.

We next note from the output in part b that

$$b = 0.12486$$

$$SE_b = 0.01290$$

For a 90% confidence interval from Table C with $n = 14$ (and $n - 2 = 12$),

$$t^* = 1.782$$

We put these pieces together to compute the 90% confidence interval.

$$b \pm t^*SE_b = 0.12486 \pm (1.782)(0.01290) = 0.12486 \pm 0.02299$$

Exercise 11.14

a) From the output given in Exercise 11.13, part b, we see that the regression equation is

Manatees killed = - 41.43 + 0.12486 Boats

Plug in the value 716 for boats (boats were measured in thousands of boats when fitting the line, so we plug in 716, not 716,000) and the prediction is

Manatees killed = - 41.43 + 0.12486(716) = 47.97

b) The value 47.97 is the same as calculated in part a. We are interested in predicting a single value (the number killed in a future year in which boat registrations are 716,000), so the appropriate interval is the 95% P.I. From the output the 95% prediction interval is thus

37.46 to 58.48

c) The number killed in 1991, 1992, and 1994 fall within the prediction interval given in part b. The number for 1993 is outside (less than) the interval. Of course, the predictions assume that conditions did not change. If measures were enacted to protect manatees, then predictions based on previous years would not apply.

Exercise 11.15

From the output in Exercise 11.14

$$SE_{\hat{\mu}} = 2.23$$

In Exercise 11.14 we also found

$\hat{y}$ = predicted number of manatees killed for 716,000 powerboats

$$= -41.43 + 0.12486(716) = 47.97$$

In addition,

$$n - 2 = 14 - 2 = 12$$

t^* for a 90% confidence interval = 1.782

The desired 90% confidence interval is therefore

$$\hat{y} \pm t^* SE_{\hat{\mu}} = 47.97 \pm 1.782(2.23) = 47.97 \pm 3.97$$

Exercise 11.18

a) Here is the desired scatterplot.

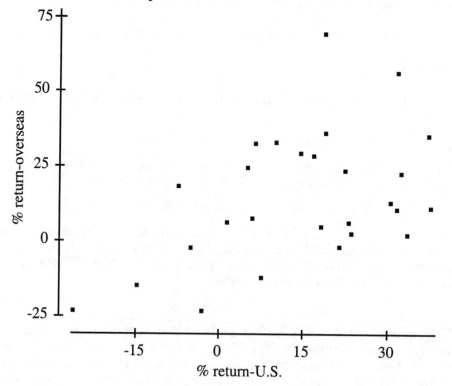

b) From the output given,

$$b = 0.6181$$

$$SE_b = 0.2369$$

hence

$$t = \frac{b}{SE_b} = 0.6181/0.2369 = 2.609$$

The sample size is $n = 27$, so

$$\text{Degrees of freedom} = n - 2 = 25$$

From Table C we estimate the P-value to be between 0.02 and 0.01 (we double the tail probabilities at the top of Table C since we have a two-sided alternative).   The evidence appears to be reasonably strong for a linear relationship between U.S. and overseas returns.

c) Based on the output in part b, the equation of the least-squares regression line is

$$\% \text{ return-overseas} = 5.683 + 0.6181 \ (\% \text{ return-U.S.})$$

If we substitute into this equation $\% \text{ return-U.S.} = 15$, we get

$$\% \text{ return-overseas} = 5.683 + 0.6181(15) = 14.9545$$

which agrees with the value given in the output.

Two 90% intervals are given in the output. Since we are predicting a single value (the value next year if U.S. stocks return 15%), we use the 90% prediction interval (P.I.). According to the output, this interval is

$$-19.65 \text{ to } 49.56$$

d) The value of r^2 is only 0.214. This indicates that only 21.4% of the variation in the return on foreign investments is explained by the least-squares regression line using the return on U.S. stocks as a predictor. Since 78.6% of the variation is unexplained, we would not expect predictions using the least-squares regression line to be very accurate, hence useful. Indeed, our 90% prediction interval is very wide and not very useful.

Exercise 11.19

a) Here is a plot of the residuals against U.S. % return.

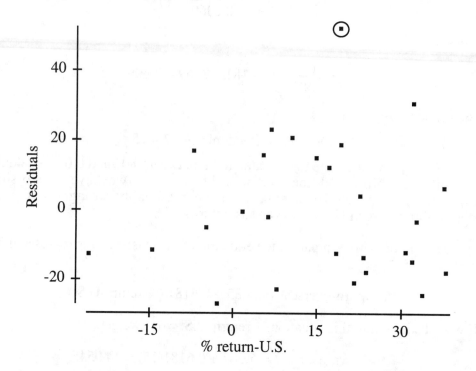

The vertical spread of the residuals appears to increase somewhat as the % return on U.S. stocks increases. This suggests a moderate violation of the assumption that σ, the standard deviation of the response, is constant for all values of the predictor variable. There may also be a slight outlier (the point circled in the plot), suggesting a moderate violation of the assumption of normality.

b) Here is a histogram of the residuals (you might also try making a stemplot).

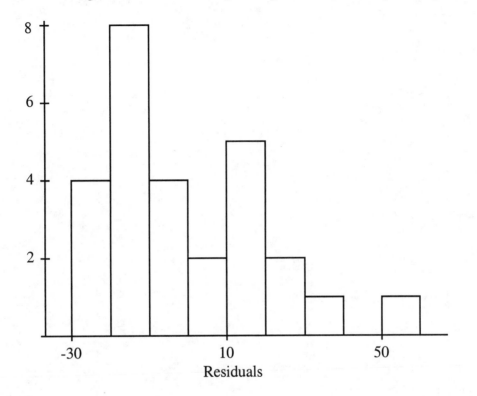

The shape is somewhat skewed to the right. We see a possible outlier, the residual with value 52.22 (circled in the scatterplot in part a) If one examines the original data in Chapter 2, this point corresponds to the year 1986 when the return on foreign stocks was unusually large.

CHAPTER 12

NONPARAMETRIC TESTS

SECTION 12.1

OVERVIEW

Many of the statistical procedures described in previous chapters assumed that the samples were drawn from normal populations. **Nonparametric tests** do not require any specific form for the distributions of the populations from which the samples were drawn. Many nonparametric tests are **rank tests**; that is, they are based on the **ranks** of the observations rather than on the observations themselves. When ranking the observations from smallest to largest, tied observations receive the average of their ranks.

The **Wilcoxon rank sum test** compares two distributions. The objective is to determine if one distribution has systematically larger values than the other. The observations are ranked, and the **Wilcoxon rank sum statistic W** is the sum of the ranks of one of the samples. The Wilcoxon rank sum test can be used in place of the **two-sample t test** when samples are small or the populations are far from normal.

Exact P-values require special tables and are produced by some statistical software. However, many statistical software packages give only approximate P-values based on a normal approximation, typically with a continuity correction employed. Many packages also make an adjustment in the normal approximation when there are ties in the ranks.

GUIDED SOLUTIONS

Exercise 12.5

KEY CONCEPTS - ranking data, two-sample problem, Wilcoxon rank sum test

a) Is this a one-sided or two-sided alternative? It is best to state the null and alternative hypotheses in words. You may wish to reread Exercise 7.37.

H_0:

H_a:

b) Order the observations from smallest to largest. Use a different color or underline the observations in the control group. This will make it easier to determine the ranks assigned to each group.

Now suppose that the first sample is the control group and the second sample is the DDT group. The choice of which sample we call the first sample and which we call the second sample is arbitrary. However, the Wilcoxon rank sum test is the sum of the ranks of the first sample, and the formulas for the mean and variance of W distinguish between the sample sizes for the first and the second samples. What are the ranks of the control observations? Use these ranks to compute the value of W.

$W =$

What are the values of n_1, n_2, and N? Use these values to evaluate the mean and standard deviation of W according to the following formulas.

$$\mu_W = \frac{n_1(N+1)}{2} =$$

$$\sigma_W = \sqrt{\frac{n_1 n_2 (N+1)}{12}} =$$

Now use the mean and standard deviation to compute the standardized rank sum statistic.

$$z = \frac{W - \mu_W}{\sigma_W} =$$

What kind of values would W have if the alternative were true? Use the normal approximation to find the approximate P-value. If you have access to software or tables to evaluate the exact P-value, compare it with the approximation.

What are your conclusions?

c) How do the results compare with those obtained using the two-sample t test in Exercise 7.37?

COMPLETE SOLUTIONS

Exercise 12.5

a) The hypotheses are

H_0: the distribution of nerve activity is the same for both groups

H_a: the nerve activity is systematically higher in either the DDT group or the control group

The alternative is two-sided, because as indicated in Exercise 7.37, the researchers did not conjecture in advance that the nerve activity would increase in rats fed DDT.

b) The observations are first ordered from smallest to largest. The observations in bold are from the DDT group.

6.642, 8.182, **8.456**, 9.351, 9.686, 11.074, 12.064, **12.207, 16.869, 20.589, 22.429, 25.050**

Suppose the first sample is the control and the second sample is the DDT group. In this case, $n_1 = n_2 = 6$, and N, the sum of the sample sizes is 12.   The control observations have ranks 1, 2, 4, 5, 6, 7 and the sum of these ranks is $W = 25$. The values for the mean and variance are

$$\mu_W = \frac{n_1(N+1)}{2} = \frac{6(12+1)}{2} = 39$$

and

$$\sigma_W = \sqrt{\frac{n_1 n_2 (N+1)}{12}} = \sqrt{\frac{(6)(6)(12+1)}{12}} = 6.25$$

and the standardized rank sum statistic W is

$$z = \frac{W - \mu_W}{\sigma_W} = \frac{25 - 39}{6.25} = -2.24$$

We expect W to be either small or large when the alternative hypothesis is true, as DDT could cause the control group to tend to receive either the smaller or larger ranks. The approximate P-value is $2 \times P(Z \leq -2.24) = 0.0250$, where we have doubled the probability since the alternative is two-sided.

If we use the continuity correction, we act as if the number 25 occupies the interval from 24.5 to 25.5. This means that to compute the P-value, we first calculate the probability that $W \leq 25.5$:

$$P(W \leq 25.5) = P\left(\frac{W - \mu_W}{\sigma_W} \leq \frac{25.5 - 39}{6.25} \right) = P(Z \leq -2.16) = 0.0154$$

and then double this giving the P-value $2 \times 0.0154 = 0.0308$. The exact P-value using tables is $2 \times 0.0130 = 0.0260$. Typically the continuity correction improves the approximation, but in this example the approximation is slightly better without the correction. The data give strong evidence that the measures of nerve activity are systematically higher in the DDT group.

c) The results aren't different enough to change the conclusions of the study.

SECTION 12.2

OVERVIEW

The **Wilcoxon signed rank test** is a nonparametric test for matched pairs. It tests the null hypothesis that there is no systematic difference between the observations within a pair against the alternative that one observation tends to be larger.

The test is based on the **Wilcoxon signed rank statistic W^+**, which provides another example of a nonparametric test using ranks. The absolute values of the differences between matched pairs of observations are ranked and the sum of the ranks of the positive (or negative) differences gives the value of W^+. The **matched pairs t test** is another alternative test for this setting.

P-values can be found from special tables of the distribution or a normal approximation to the distribution of W^+. Some software computes the exact P-value and other software uses the normal approximation, typically with a ties correction. Many packages make an adjustment in the normal approximation when there are ties in the ranks.

GUIDED SOLUTIONS

Exercise 12.17

KEY CONCEPTS - matched pairs, Wilcoxon signed rank statistic

First give the null and alternative hypotheses

H_0:

H_a:

To compute the Wilcoxon signed rank statistic, first order the absolute values of the differences and rank them. Due to the large number of ties in this exercise, be careful when computing the ranks. Any tied group of observations should receive the average rank for the group. (The negative observations are in bold and italics.)

Absolute values	Ranks
1	1.5
1	1.5
3	4.5
3	4.5
3	4.5
3	4.5
4	7.5
4	7.5
7	9
10	10.5
10	10.5
12	12
22	13
31	14

To see how the ranks are computed, the 1's would get ranks 1 and 2, so their average rank is 1.5. The 3's would get ranks 3, 4, 5, and 6, so their average rank is 4.5, and so on. If W^+ is the sum of the ranks of the positive observations, compute

$W^+ =$

Evaluate the mean and standard deviation of W^+ according to the formulas.

$$\mu_{w^+} = \frac{n(n+1)}{4}$$

$$\sigma_{w^+} = \sqrt{\frac{n(n+1)(2n+1)}{24}}$$

Now use the mean and standard deviation to compute the standardized rank sum statistic.

$$z = \frac{W^+ - \mu_{w^+}}{\sigma_{w^+}} =$$

Do you expect W^+ to be small or large if the alternative is true? Use the normal approximation to find the approximate P-value.

What are your conclusions?

COMPLETE SOLUTIONS

Exercise 12.17

The null and alternative hypotheses are

H_0: healing rates have the same distribution for the both groups

H_a: healing rates are systematically lower in the experimental group

The Wilcoxon signed rank statistic is

$$W^+ = 4.5 + 7.5 + 10.5 = 22.5$$

The values for the mean and variance are

$$\mu_{w^+} = \frac{n(n+1)}{4} = \frac{14(14+1)}{4} = 52.5$$

and

$$\sigma_{w^+} = \sqrt{\frac{n(n+1)(2n+1)}{24}} = \sqrt{\frac{(14)(15)(28+1)}{24}} = 15.93$$

and the standardized signed rank statistic $\dot{W}^+$ is

$$\frac{W^+ - \mu_{w^+}}{\sigma_{w^+}} \le \frac{22.5 - 52.5}{15.93} = -1.88$$

If changing the electric field reduces the healing rate, we would expect the differences (experimental - control) to be negative. Thus the ranks of the positive observations should be small and we would expect the value of the W^+ to be small when the alternative hypothesis is true. The approximate P-value is $P(Z \le -1.88) = 0.0301$.

The output from the Minitab computer package gives a similar result. Many computer packages, including Minitab, include a correction to the standard deviation in the normal approximation to account for the ties in the ranks.

```
Wilcoxon Signed Rank Test

TEST OF MEDIAN = 0.000000 VERSUS MEDIAN L.T. 0.000000

                N FOR    WILCOXON
          N     TEST     STATISTIC   P-VALUE
Nerve     14     14        22.5       0.032
```

The data show that changing the electric field reduces the healing rate.

SECTION 12.3

OVERVIEW

The **Kruskal-Wallis test** is the nonparametric test for the **one-way analysis of variance** setting. In comparing several populations, it tests the null hypothesis that the distribution of the response variable is the same in all groups and the alternative hypothesis that some groups have distributions of the response variable that are systematically larger than others.

The **Kruskal-Wallis statistic** H compares the average ranks received for the different samples. If the alternative is true, some of these ranks should be larger than others. Computationally, it essentially arises from performing the usual one-way ANOVA to the ranks of the observations rather than the observations themselves.

P-values can be found from special tables of the distribution or a chi-square approximation to the distribution of *H*. When the sample sizes are not too small, the distribution of *H* for comparing *I* populations has approximately a chi-square distribution with *I* - 1 degrees of freedom. Some software computes the exact *P*-value and other software uses the chi-square approximation, typically with an adjustment in the chi-square approximation when there are ties in the ranks.

GUIDED SOLUTIONS

Exercise 12.21

KEY CONCEPTS - one-way ANOVA, Kruskal-Wallis statistic

a) Think carefully about the differences in the hypotheses being tested.

ANOVA test

H_0:

H_a:

Kruskal-Wallis test

H_0:

H_a:

b) Find the median for each group. Recall that the median is the middle observation after the observations have been ordered. When the sample sizes are even, the median is the average of the two middle observations.

Nematodes	Median
0	
1000	
5000	
10000	

Using the information in the medians, do the nematodes appear to retard growth?

To compute the Kruskal-Wallis test statistic, the 16 observations are first arranged in increasing order. That step has been carried out, where we have kept track of the group for each observation. Fill in the ranks. Remember that there is one tied observation.

Growth	3.2	4.6	5.0	5.3	5.4	5.8	7.4
Group	10000	5000	5000	10000	5000	10000	5000
Rank							

Growth	7.5	8.2	9.1	9.2	10.8	11.1	11.1
Group	10000	1000	0	0	0	1000	1000
Rank							

Growth	11.3	13.5
Group	1000	0
Rank		

Now fill in the table, which gives the ranks for each of the nematode groups and the sum of ranks for each group.

Nematodes	Ranks	Sum of Ranks
0		
1000		
5000		
10000		

Use the sum of ranks for the four groups to evaluate the Kruskal-Wallis statistic. What are the numerical values of the n_i and N in the formula?

$$H = \frac{12}{N(N+1)} \sum \frac{R_i^2}{n_i} - 3(N+1) \quad =$$

The value of H is compared with critical values in Table E for a chi-square distribution with $I - 1$ degrees of freedom, where I is the number of groups. What is the P-value and what do you conclude?

COMPLETE SOLUTIONS

Exercise 12.21

a) The null and alternative hypotheses for the ANOVA test are

$$H_0: \mu_0 = \mu_{1000} = \mu_{5000} = \mu_{10000}$$

$$H_a: \text{not all four means are equal}$$

and the null and alternative hypotheses for the Kruskal-Wallis test are

$$H_0: \text{seedling growths have the same distribution in all groups}$$

$$H_a: \text{seedling growth is systematically higher in some groups than in others}$$

When the distributions have the same shape, the null hypothesis for the Kruskal-Wallis is that the median growth in all groups are equal, and the alternative hypothesis is that not all four medians are equal.

b) The medians for the nematode groups are given in the table. The ordered observations from the first group are 9.1, 9.2, 10.8, 13.5. The median is the average of the two middle observations, $(9.2 + 10.8)/2 = 10.0$.

Nematodes	Median
0	10.00
1000	11.10
5000	5.20
10000	5.55

The medians suggest that nematodes reduce growth rate. There appears not to be a decrease when nematodes are at the level of 1000, but the growth drops off at 5000 and no further reduction is apparent at 10000.

It is important to note that neither the ANOVA test nor the Kruskal-Wallis test are designed specifically for the alternative that nematodes retard growth (this would be the analog of a one-sided alternative). Both procedures test the alternative of *any* differences between the groups.

The computations required for the Kruskal-Wallis test statistic are summarized in the tables.

Growth	3.2	4.6	5.0	5.3	5.4	5.8	7.4
Group	10000	5000	5000	10000	5000	10000	5000
Rank	1	2	3	4	5	6	7

```
Growth   7.5    8.2    9.1    9.2   10.8   11.1   11.1
Group  10000   1000     0      0      0   1000   1000
Rank       8      9     10     11     12   13.5   13.5
```

```
Growth  11.3   13.5
Group   1000      0
Rank      15     16
```

Nematodes	Ranks	Sum of Ranks
0	10, 11, 12, 16	49
1000	9, 13.5, 13.5, 15	51
5000	2, 3, 5, 7	17
10000	1, 4, 6, 8	19

$$H = \frac{12}{N(N+1)} \sum \frac{R_i^2}{n_i} - 3(N+1) = \frac{12}{16(16+1)} \left(\frac{49^2}{4} + \frac{51^2}{4} + \frac{17^2}{4} + \frac{19^2}{4} \right) - 3(16+1)$$

$$= 11.34$$

Since $I = 4$ groups, the sampling distribution of H is approximately chi-square with $4 - 1 = 3$ degrees of freedom. From Table E we see that the P-value is approximately 0.01. There is strong evidence of a difference in seedling growth between the four groups.

The Minitab software gives the following output when doing the Kruskal-Wallis test. The medians, average ranks (in place of sums of ranks), H statistic, and P-value are given. The H statistic with an adjustment for ties in the ranks is also given.

```
LEVEL      NOBS     MEDIAN   AVE. RANK
   1         4      10.000      12.3
   2         4      11.100      12.8
   3         4       5.200       4.2
   4         4       5.550       4.7
OVERALL     16                   8.5
```

```
H = 11.34   d.f. = 3   p = 0.010
H = 11.35   d.f. = 3   p = 0.010 (adjusted for ties)
```